DITTRICH

FUTURE = EARTH?!

A Solution for Climate Change and World Peace

Future=Earth?!
A Solution to Climate Change and World Peace

1. edition

Publisher: Magyar Klímavédelmi Ltd.
8. street Hegedűs János Pécs, Hungary H-7630

Author: Dr. Ernő Dittrich
web: www.justdobetterworld.com
blog: https://justdobetterworld.blogspot.com/
e-mail: justdobetterworld@gmail.com

Professional proofreaders:
Dr. Ferenc Szilágyi *honorary professor*
Judit Pap *psychologist-kinesiologist*
Proofreader: robertjamesrobson@gmail.com
Cover design: Attila Márok // www.weiser.hu
Design & layout: Zoltán Novreczky // www.hangtars.hu

Print finishing: Molnár Nyomda és Kiadó Ltd.
28. street Légszeszgyár Pécs, Hungary H-7622
iroda@nyomdamolnar.hu

© Dr. Ernő Dittrich – All rights reserved!

ISBN 978-615-01-7802-8

*„Do it now!
Sometimes later
becomes never."*

DR. SANJAY TOLANI
economic consultant and coach

This book is independent from any political or religious organizations and from any ideological or other groups. The spirit of this book considers every living entity as equal; this way, it is obviously unnecessary to belong to any group. The only and selfless aim of this book is to get Humankind out of its climate change crisis. This principle itself makes it obligatory for the book not to belong to any economic or other interest groups, which would otherwise prevent the book from being objective and useful.

The Author

Content

THE CRISIS OF RECYCLED PAPER 11

CHAPTER 1: HOW SERIOUS IS THE PROBLEM - OR WHAT WILL OUR FUTURE BE LIKE? 15

1.1. The realistic future caused by climate change 15
1.2. Why climate change cannot be stopped by
the current methods 21
    1.2.1. The physical reason and some basic definitions to begin ... 21
    1.2.2. The Plimsoll line effect and the tipping points - or
why we need to keep global warming under 1.5-2 °C?! 22
    1.2.3. The economic reason 27
    1.2.4. The biological, ecological reason 29
    1.2.5. Habitat and vegetation fragmentation, soil degradation,
biodiversity reduction 30
1.3. The vision - or the realistic future as I see it 32
1.4. The overwhelming Adam Smith vision - or
how we could get so far 34
1.5. The dividing of our tasks between generations 38

CHAPTER 2: PATTERNS OF THOUGHT NEEDED FOR A SOLUTION 41

2.1. What is the second step? 41
2.2. The mistaken paradigm of competition 45
2.3. Every Life is equal 48
2.4. The new level of evolution 49
2.5. To what extent are you a Life supporter? 52
2.6. The respect of Nature and happiness 56
2.7. The common root of personal unhappiness and
climate change 57
2.8. The manifestation 59
2.9. Your thinking covers your body's wisdom 64
2.10. Quantifiable happiness - or the levels of consciousness 67

2.11. Group dynamics and some conclusions 81
2.12. The ignored evolution.. 84
2.13. Rationality versus heart-coherent objectivism - or
       the exaggerated rational approach 90
2.14. What is happiness? Can we search for
       happiness consciously? 97

CHAPTER 3:
THE SOLUTION: THE 6 PROGRAMS OF CHANGE 101

CHAPTER 4: THE FIRST PROGRAM OF CHANGE:
THE PROGRAM OF REVITALIZATION 103
4.1. Is the whole world really possessed by humans?..................... 103
4.2. The protection of natural areas 104
4.3. The main steps of the Program of Revitalization 105
4.4. Rules in connection with natural areas, the Natural Territorial
     Development Directive, revitalization 106
4.5. Providing for the protection of natural areas.......................... 108
4.6. Social rudiments 110
4.7. The benefits of the Program of Revitalization......................... 111

CHAPTER 5: THE SECOND PROGRAM OF CHANGE:
THE PROGRAM OF AGGLOMERATION 115
5.1. Targets, basic principles 115
5.2. The principle of interdependence 120
5.3. The Global Grid 123
5.4. Balance strategies 127
5.5. The power of bottom-up initiatives - or
     how is this system established? 131
5.6. Land use ratio and the modification of
     agglomeration boundary lines... 133
5.7. Traveling and trade among agglomerations 133
5.8. The Global Happiness Index (GHI) 135
5.9. Territories requiring international protection 139
5.10. The effects of introduction 139

5.11. Frequently asked questions, counterarguments 141

CHAPTER 6: THE THIRD PROGRAM OF CHANGE: THE PROGRAM OF POPULATION 145
6.1. Oops! We are overpopulated 145
6.2. How can we solve overpopulation democratically? 147
6.3. Society's phobia of aging 151
6.4. Religions and contraception 153
6.5. Families and childcare hardships 154
6.6. The main effects of the program 155

CHAPTER 7: THE FOURTH PROGRAM OF CHANGE: THE PROGRAM OF HAPPINESS 157
7.1. Selflessness, community Life and the GHI 157
7.2. The Media Ethic Codex 158
7.3. The method of Life-supportive balance 164
7.4. Introducing the general mental hygiene system and
     social prevention: the practical measure of GHI 165
7.5. Kinesiological screening of important professions 170
7.6. Restructuring the educational and child care systems 170
7.7. The mission system among agglomerations 173

CHAPTER 8: THE FIFTH PROGRAM OF CHANGE: THE PROGRAM OF SOCIETY 175
8.1. The future society 175
8.2. What is the difference between the European Union and
     the union of agglomerations? 178
8.3. The long-awaited world peace is close 178
8.4. Global innovation, research and the space program 180
8.5. The sun society 181
8.6. The eating habits of the future society 182
8.7. The transition to a service-based society 183
8.8. Labeling obligation for products and services 184
8.9. A common world language 187
8.10. White is the winner - the albedo trick 187
8.11. Group meditation 188

8.12. Some incorrect thinking processes.. 189

CHAPTER 9: THE SIXTH PROGRAM OF CHANGE: THE PROGRAM OF ECONOMY 193
9.1. GHI instead of GDP 194
9.2. The companies above humanity 195
9.3. Profit vs a moneyless world.. 196
9.4. Deleting the principle of limitless growth 198
9.5. Tax relief and interest rebate 199
9.6. Taxing of derelict land 199
9.7. Environmental Impact-, People Harming- and Labeling Taxes 201
9.8. Mitigating economic inequalities through agglomeration rules 203
9.9. Solidarity Tax - the principle of direct and indirect emission (breaking the lobby power of environment-damaging industries) 204
9.10. CO_2 extraction from the atmosphere 205
9.11. The switch to regenerative agriculture....................................206
9.12. The Global Service List and economic restructuring..............206
9.13. The gradual disappearance of competition 209
9.14. The circular economy, zero-waste future................................209

CHAPTER 10: THE INTERACTION OF THE 6 PROGRAMS; TRANSITION TO ANOTHER DIRECTION OF DEVELOPMENT 211

CHAPTER 11: THE INDIVIDUAL'S ROLE............................217
11.1. The four main principles of a carbon-neutral life......................217
11.2. The power of an individual's decisions and example, or the principle of impact multiplication..219
11.3. Your own personal happiness is the most important! 220
11.4. Everything that boosts your selflessness saves the world....... 221
11.5. There is a huge potential in personal lifestyle change 222
    11.5.1. You can save the Earth with your eating habits!..............222
    11.5.2. Travel closer. Fly only if it is absolutely necessary!........223

11.5.3. Do not smoke (so much)! The example of conscious thinking according to different life cycles 225
11.5.4. Three simple habits to reduce your CO_2 emission by more than 10% 227
11.5.5. Did you ever imagine that composting could be so useful? 230
11.5.6. Five more eating habits to reduce your impact on climate change by 10 to 25% 231
11.5.7. Change to electricity wherever possible! 236
11.6. Having and raising children responsibly 240
11.7. Vote responsibly! 240
11.8. The principle of interdependence on the individual level 240
11.9. The endless possibilities of technical solutions 240
11.10. Protect, take care of, respect and support Nature 242
11.11. The power of individual and group meditation 242
11.12. More than a hundred practical possibilities 242

CHAPTER 12: THE IMPLEMENTING STEPS OF THE 6 PROGRAMS IN THE SHORT, MEDIUM AND LONG TERMS 249

CLOSING REMARKS 257

ACKNOWLEDGEMENT 259

LITERATURE 261

HIGHLY RECOMMENDED MOVIES ABOUT CLIMATE CHANGE 264

APPENDICES 265

APPENDIX 1: MORE ABOUT EGO 267
Observing the (your) ego 267
(Your) ego's features: separation 269

Why is the ego-free person more efficient, healthier and happier than the one with a strong ego - or the nine tricks of (your) ego 272

APPENDIX 2: MORE ABOUT GAMES 281

APPENDIX 3: MORE ABOUT THE LEVELS OF CONSCIOUSNESS 285
Shame and guilt 285
Apathy and grief 286
Fear 288
Desire 290
Anger 292
Pride 294
Courage 298
Neutrality 301
Willingness 303
Acceptance 306
Reason 308
Love 310
Joy and peace 312
Enlightenment 315

The crisis of recycled paper

This book is made of 100% recycled paper. But have you ever realized that it's much more complicated to find a publisher who will agree to publish a book in this way? What's more, the production costs are about 30% higher. It's weird, isn't it? We live in an abnormal world. If we cut down natural forests (for free), make paper from them, then make books from that paper, it's cheaper than just reusing waste paper. If an alien with high intelligence was watching this as an outsider, they would be astonished to see us. Destroying Life is cheaper than protecting it? How can this be possible? Indeed, it is so—at least according to our current and 'modern' economic system.

All through my childhood I could not understand how humans could live this way, and how it could be good for them.[1] As I learned more and more about the world, I fostered the thought that things should be changed. The competition between people, the fight for the seemingly few resources for which they drag each other through the mud, the wanton fight of egos and the irresponsible destruction of the environment have all been senseless and incomprehensible to me ever since then.[2] As time passed, that puzzled little child became an experienced expert who has delivered his integrated knowledge in this book to enable you to shape this knowledge according to your life, develop it further, and form it in keeping with your own thoughts. I do not want to tell you what to do and what not to do; I want to show you a framework that you can reshape as you like. Please add what you like and ignore what you cannot use. The main aim is that the thoughts generated by this book should be realized in actions, and that more of us should act for the cause of good.

Being an engineer, the solution-focused mindset which permeates this book is crucial to me. Through decades spent searching, reflecting and thinking, I have always sought the answer to my question of how

---

[1] I am going to write the words Humankind, Life and Nature deliberately with capital letters in the book, showing my respect to these three wonderful entities.
[2] You can read about the nature of ego in Appendix 1 if you are interested in it in details.

we could set the development of Humankind, living in such a wrong way due to its old and bad habits, in a good direction? I can see the difficulties of the fight against these bad habits as I have to fight with my own day by day to be able to live in a more environmentally aware way and to be more accepting. It requires gradual and mindful self-development within my own life as well as by everyone else. For instance, a serious foible of mine is that I drive really fast, using 30% more energy for transport than if I drove more calmly and slowly. And I could come up with plenty of other items for improvement on my to-do list. Currently, I am working on being a more active participant of the fight against climate protection even in that field. There is always something you can improve on; you can always develop. I am convinced that this book offers the problem-solving system for climate change, which as you will see is also a remedy for the unrest of the world and for many other social problems. As every social problem can be healed at the individual level, I am quite sure that this system will affect your personal life and happiness in a pleasant way. Moreover, I make no secret that my aim is to pick fights with destructive social habits, and with all of the destructive aspects of the anti-climate socio-economic and social systems. The ideas within the book might seem divisive, but no change can be imagined unless we have the courage to look into the mirror.

Every objective realization starts with a thought, so that is the most difficult point to reach. Once the right package of thoughts is clarified, there's a good chance that everything will go in a better direction. The honest and pure desire of my heart is that Humankind should continue its future in a way that will create a happy and productive future for the coming generations.

This book is firstly a vision, secondly a guide, thirdly a study to prepare a framework directive. I would therefore like to ask you, dear reader, not to reject the visions about the future with such instinctive reactions as 'That's impossible! People today are too selfish to do that!' or 'We can't do things that way because there isn't enough economic power in society for it to succeed!' and so on. I would like to ask you to read the book and consciously hold back every prejudice until the end.

The book is going to show you many patterns of thought that differ from current ones, as this is the key to the solution. With Humankind's current approach and mindset, we will not last long. It can be stated that the thinking mechanisms currently ruling in the western world are incapable of leading the human race into a happy and peaceful future. I therefore respectfully ask you to stay open while reading the book even if at first it may seem illogical in places. It is always the statements that seem weird at first that will start the biggest change within ourselves, since they set our way of thinking in a new direction. That is why we are averse to them at first, then they mature slowly in our minds, and later they integrate into our lives. It is a natural human reaction; everybody works like this. Beyond being open, I respectfully ask you one more thing: supposedly questions and concerns will be arising continuously as you read the book. Please remember these and set them aside. I can promise that you will have found all the answers to your questions by the end of the book.

    I am asking you to let the big picture depicted by this book be built up within yourself, and ignore the obstacles that come up instinctively in your mind. These are the natural reactions of the old and traditional way of thinking. If the whole picture of my vision is built up by the end of the book, and you still have concerns even considering that there are a few generations worth of time for its realization, please write them to me, as every piece of constructive criticism helps this system to develop further. You can find my e-mail address at the beginning of the book. Of course, we are happy to get comments and suggestions on our website and on my blog; these are also listed at the beginning of the book. I thank you in advance if you read this book and help this world-saving system with your opinion, with which you also help us to do more to set social changes in the right direction. Besides this book, I would like to serve our future with a lot of free content, advanced training and subsequent books about which you can find further information on the website noted at the beginning of the book. As I believe in teamwork, if you would like to contribute to the development, distribution and integration of the system, I am happy to welcome you into our community. To stop climate change, cooperation is needed!

CHAPTER 1

How serious is the problem, or: what will our future be like?

1.1. The realistic future caused by climate change

Imagine the past, present and future of Humankind—namely our whole history—as a long journey. Our level of development in respect to the given time periods is characterized by the quality and width of the road. In prehistoric times this road was just a path, very curved and narrow, crossed by herds of animals. On this dangerous path, fallen trees, mud, wallows and in places gullies made it more difficult to move forward. In ancient times this path widened; short sections of the path were even paved. In places, carts and horses could go on it and humans could proceed at a relatively quicker pace in its development. In the Middle Ages this road sometimes became a wallow, or narrowed then widened again, but it did not change considerably. Moving forward on the road was monotonous and hard. Then came modern times when this road widened out suddenly. Now carts and carriages traveled on it, and most of the people went along with their heads held high. Our road was reshaped into a multilane thoroughfare. During the two world wars, minefields and bomb craters made it difficult to move forward on the road, making Humankind's development torturous and extremely difficult. After this came the postwar period when within only a few decades the quality of the road suddenly improved a lot: first concrete, then asphalt roads were built. These roads became wider and wider and of better and better quality, and faster and faster vehicles turned up on them. Today, the road that characterizes the development of our civilization is an amazingly wide motorway where we can rush speedily with advanced vehicles on high quality paving (unfortunately, we often do it wantonly, sitting alone in our multi-person cars). Humankind has experienced such technological and scientific development in the last decades that had never been seen before, and the average standard of living has never been as high in our history as it is today. Life in the western world is characterized by relative public safety, high-level health care, well-being, food safety, and comfort. We can be grateful to Life that we are on this part of the Earth and in this era, since a lot of our peers and our ancestors, looking back for centuries and millennia, have never lived as well as we do—at least from a materialistic aspect.

If we continued this road in our imagination into the future, this really wide and high-level motorway system that Humankind is moving on, it would suddenly start to narrow down. The lanes of the motorway would be reduced, while the traffic would not. There would be neither time nor capacity to rebuild them, so this road, going into the future, would have poorer and poorer paving. Transportation would slow down and everybody would blame others for not being able to proceed properly. Our future would be stuck in a horrible and unsolvable traffic jam.

If we continued our collective life in this social, economic and political way, the time would come when this road became a one-lane road again. Afterwards, the solid pavement would disappear, and Humankind would wallow in the mud barefoot again. Moreover, considering current tendencies, there is a genuine chance that we would fall back into a social and economic state equal to prehistoric and ancient times' hard periods—or worse. I am not saying this because I am a Nostradamus type of prophet, but rather because the current world models, scientific results and facts about the world as we know them see us heading in this direction.

If we continue living the way we are living now, the climate in the rest of the Amazon's jungles would become similar to the savannah's climate in a short time, and this would result in the dying out of the remaining jungles, accelerating the climate change in the world. The other factor that generates huge changes is the melting of the ice caps, through which the albedo[3] of the globe would change dramatically, also greatly accelerating global warming. In addition, the levels of the oceans and seas would increase, threatening the housing and lives of hundreds of millions of people. Meanwhile, due to overfishing, the oceans of the world would become ocean-deserts with no high-level lifeforms. Because of the climate change and the extremely rapid soil degradation, agriculture would be able to produce less and less food year by year, and as a result the remaining natural lands would be broken up to become farmland. This would result in the collapse of our natural ecological systems, which have crucial roles in maintaining the climate system everywhere in the world, thus speeding up the climate change even more. The result would be that agriculture would not be able to produce food for even 10% of the world's population. Meanwhile, the melting of the ice caps would stop the main currents in the oceans, because of which the climate would change in every region incredibly fast, reaching the point within a few years where the remaining wildlife could not adapt in time. Everything that had stayed alive to that point would die out, independent of whether they are parts of the natural ecosystem or crops. This process would be hastened up by the thawing of permafrost[4], under which an incredible amount of methane[5] is piled up. If all that methane is released into the atmosphere, literally all hell would break loose, as the warming would speed up even more rapidly. Humans would lose control over the destructive powers of Nature forever. The final mass extinction of the remaining wildlife would occur, since they would not be able to adapt to the accelerating changes.

Thus it seems we have stepped onto a road that can lead to a huge catastrophe, the likes of which have not been seen since the extinction of the dinosaurs. Scientists have named it the Sixth Extinction Event, which has technically already begun. The problem is that it is not being caused by a meteor, but by Humankind.

---

[3] albedo: the ratio of light reflected back from the surface of the Earth from the amount of light reached its surface
[4] broad areas with frozen surface mainly in taiga areas
[5] about 23 times more powerful a greenhouse gas than CO_2

Due to the processes mentioned above, famine would be continuously increasing, people would migrate in bigger and bigger groups to the places where there is still chance for a life worthy of human living. It would generate increasing migration that would cause huge social tensions as more and more people would increasingly fight for the remaining resources. As a result, the social systems would break up, the economy would collapse, and health care could not be maintained. More and more epidemics and diseases would decimate the population, not to mention the increasing famine on top of that. Due to the migration, epidemics, famine, and all the social tensions coming from these, the state system would also collapse, resulting in armed groups fighting against each other for the remaining resources, reducing the number of people. The sad end result of all these happenings is that the enormous human population, which has been increasing at its maximum for some decades, would suddenly start decreasing. Because of the increasing climate change, we would face a dry, extinct planet with almost no wildlife and full of dust storms—a situation that we could not change even with all our technical and scientific knowledge. The end result would be not only the extinction of humans from the Earth: 99.99% of species would also be dragged down and killed by our neglectfulness. Thus Humankind, the developed, proud and supposedly erudite human species, will have generated the whole extinction of Earth essentially by itself. We are like a well-organized suicide squad which **is being taken to destruction by our current social-economic structures.**

When most of us think of this problem set, we think 'Okay, but that's the distant future! We will have time to react by then! The politicians and scientists will definitely figure something out!' The problem is that the beginning of the big destruction is said to be in about 2040, but if we look behind the words, it has already begun. The symptoms can be hidden today by technology and the economy. But from about 2040, famine, migration, the breakdown of economic and health care systems would be typical in rapidly increasing numbers of areas, and even more social systems would become more anarchic. Of course, everything would begin in the poorer countries but would gradually reach everyone. There would be such great pressure on those few countries which could still operate due to the burden of solidarity-based aid and the reception of migrants that they would collapse. In about 2080 we would be living in an extinct world, hit with dust storms, where 90% of humans would have already died out. The remaining people would live their everyday lives at a very low technical level, continuously starving and being afraid of some anarchic violent organization taking their few remaining resources.

The date of 2040 is horribly close, as it was already 2021 when this book was being written. This is a mere 19 years altogether: 19 summers and 19 winters. It is an incredibly short time: our generation is going to be old when we experience it, but our children are going to be in middle aged and our grandchildren in their childhoods. I do not wish this horrific future for my children or future grandchildren either! It would be terrible to look into the mirror with the feeling that I did not do everything to change this future. It is terrible to think that in 2040 my youngest son will celebrate his 34th birthday and be theoretically at the beginning of the most beautiful period of manhood. How could he have children in a world like that? At the same time, imagine that if the coronavirus could cause such a mess in society in 2019, what effects would the above-mentioned consequences of climate change have on us? The coronavirus is just a small breeze compared to the storm approaching us. But we, instead of fleeing from the storm, are running with increasing speed in its direction, like a suicide attacker to whom it does not matter what the final outcome will be. The whole human race does this unconsciously and irresponsibly. Unfortunately, although people are really smart on the individual level, as the global level of education has never before been this high, collectively we seem more stupid than most of the animal species. The gap between the sense of an individual and the social sense is really huge. You will soon see and understand clearly, dear reader, how our stuffy and narrow-minded social habits are leading our lives.

The bad news is that if we do not change our political, economic, social and technical systems and our way of Life, this will be our future. It is already sure. There can be a maximum of one or two decades' difference between the various scenarios, so it could happen that the events predicted by 2040 would happen 'only' by 2060. The final extinction might happen in 2100, but we cannot hope for dramatic changes if we do not change our present.

The path of the future is narrowed down. The line that can take us to survival has become really thin. We have to take action before it vanishes under our feet!

If you think everything that I have written is an overstatement, unfortunately I have bad news for you: the climate projections for 2020 as put together by scientists 20-30 years ago are happening exactly as predicted, and all the events are following the scenarios of the worst-case models. We have done almost nothing to halt these processes. The scientific models had not even properly taken into consideration that some unfavorable effects would speed up the processes, mutually reinforcing each other, **so the need to do something immediately is not a joke. Humankind has never been confronted by a harder or tougher fact than this in its history!**

The time we have is so short that we have to change immediately. We cannot leave it till tomorrow, next week or next year—we have to do everything we can right now. **We can change the future with 7.9 billion people's individual decisions, which may seem tiny taken alone.** Today, those 7.9 billion people's bad decisions, which may seem tiny, are causing the direction we are heading in. Thus the **"many a mickle makes a muckle"** theory does work; we can manage climate change with this theory among others. If the world goes in the direction of self-destruction, why couldn't it go in the opposite way?

The good news is that we are able to alter this horrible future if we start acting in time, individually or together, keeping the common aim in mind. The human race and the natural ecological systems of Earth still have a chance to avoid this catastrophic future!

This book is about how we can do that. Climate change is the biggest challenge in the history of Humankind. All the reading, research, work and continuous thinking regarding this topic I have done over the last 30 years have led me to develop a system which I am convinced can lead us out of this downward spiral.

But first, please recognize that **everything which is happening at the moment is our due to our irresponsibility, so it is also our responsibility where our fate will go.** Unfortunately, climate change cannot be stopped unless we change our direction.

1.2. Why can't climate change be stopped in the current way?

1.2.1. The physical reason and some basic definitions to begin with

The greenhouse gases released into the atmosphere today will have an effect for a long time because they clear from the atmosphere slowly. If we take the ratio of emissions originated by people as a basis, then it can be scientifically stated that our current emissions[6] will warm the atmosphere of the Earth for another 115-130 years on average. The following list shows how long some greenhouse gases are present in the atmosphere:
- CO remains present in the atmosphere for 50-200 years
- NO_2 remains present in the atmosphere for 50-200 years
- CH_2 remains present in the atmosphere for 12 years
- Freons remain present in the atmosphere for 65-130 years

This means that if Humankind stopped all of its emissions right at this moment, Earth would continue warming for 115-130 years. This is one of the reasons why it is not enough to reduce greenhouse gas (GHG from here on) emissions; the artificial accelerating of their extraction from the atmosphere has to be started immediately. Several technical initiatives have started recently for that reason, developing into a separate industry. Unfortunately, we are still at the very beginning of

---
[6] emission is every kind of human output that releases a given amount of harmful substances within a given time into Nature

these developments, and society does not currently provide enough resources for their expansion. These technologies are called CCU and CCS (abbreviations of the phrases Carbon Capture and Utility, and Carbon Capture and Storage). These two phrases illustrate the two main ways of development. CCU technologies consider GHG extracted from the atmosphere as a raw material and they put it back into the economy, which suits the circular economy theory. The CCS technologies store the extracted GHG in safe storage places.

1.2.2. The Plimsoll line effect and the tipping points, or why we need to keep global warming under 1.5–2 °C?!

The ecological and the climate system of our Earth is an equilibrium process. This means that if there is some outside influence, the system pursues to reestablish the state of equilibrium. As the human species developed under this equilibrium, it is in our interest is to maintain it. Unfortunately, the signs of global climate change clearly show that we have thrown our climate system off its balance artificially. It is so serious that we have exceeded by two times the maximum natural warming and cooling cycles of the past 400,000 years. The warming generated by human effects is already much more significant than those experienced on Earth during the natural warming cycles during the important part of the past evolution. This can be seen clearly in the graph below where the red 'pinpoint' shows the influence of humans.[7]

Why is this really serious for the human race? Why has Humankind stepped on thin ice? To understand the answer, I will use a metaphor.

The Plimsoll line is the level on boats under which one side of the boat, when tipped over in the wind, will not allow the boat to settle back into its original position; in other words, the boat sinks. A boat that tips because of wind or waves without going under this line always settles

---

[7] the blue line shows that during 417,000 years the CO_2 concentration of the atmosphere never went above 300 ppm, but as it is seen on the red line, it has been increasing dramatically since the 1800s

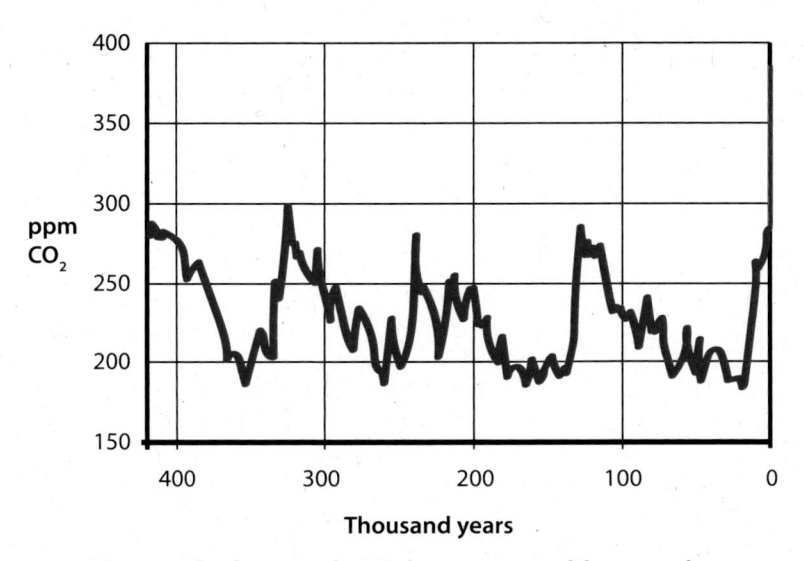

Figure 1: The change in the CO₂ concentration of the atmosphere
in the last 417,000 years
(Figure made by: I, Beroesz, CC BY-SA 3.0
https://commons.wikimedia.org/w/index.php?curid=2368657)

back into its original position; that is to say, the boat finds its balance after a while. So, from that viewpoint, a boat floating on the sea is also an equilibrium system.[8]

The climate system of Earth can be illustrated with this boat metaphor. If the global climate system tips over, until a certain boundary level it will always strive to get back to its equilibrium state. A good example of this is the natural warming-cooling cycle of Earth's climate. (See the changes shown by the blue line in Figure 1.) But if the system goes over the climate's imaginary Plimsoll line, Earth's climate would search for a new state of equilibrium.

---

8   However philosophical this idea may seem, it is necessary to point out that there has never been balance on Earth and in the universe unless we consider only one short period of time. Every system like this starts somewhere and proceeds to somewhere, and it is equally true for the development of the Earth and Life. The whole Universe therefore develops (changes) through supposed states of equilibrium. This is the mindset we have distanced ourselves from in the last 150 years. We destroyed the supposed equilibrium process that in fact was the natural development and change of the Earth. When I talk about balance in the remainder of the book, I mean the natural process of Earth's changes that may seem to be in equilibrium when observed on a human scale.

The main problem is that if once we have pushed the climate off its balance, we would not know what that new equilibrium would be like. However, the chance is extremely small that the new system would be suitable for maintaining human life, which is to say that Humankind is proceeding very quickly in the direction of its own extinction. As a matter of fact, it is certain that we are going to exceed this limit within 20 to 30 years if we do not change immediately. Global climate change is a fact, and it is also a fact that we people cause it.

The Paris Climate Accord of 2015 was signed by the countries with the aim to keep global warming under a maximum of 2 °C as compared to pre-industrialization levels. This goal was set because if the average temperature goes above 2 °C, the climate will degrade to the point where it cannot be set back to its original balance. In fact, keeping global warming below 1.5 °C would be safe, but we can already see that we cannot meet this goal. The most optimistic climate models predict 1.4 °C warming in the following 100 years, while the most pessimistic ones predict 5.4 °C (IPCC 2000). It can be clearly seen that the odds are not good. The most optimistic climate models are based on instant, global cooperation and action by the whole of Humankind. Unfortunately, we are quite far from this, although this book has been written to help us all get there.

What is certain is that we are in really big trouble; there is no time to delay or postpone measures. It is every person's moral responsibility to implement the actions in their lives which they are able to perform. No one can expect more than what we are able to do.

It is therefore reasonable to ask how far we can go with this 'tipping over' situation regarding our climate. Where is the limit from which there is no turning back? Scientists found the answer to this question almost 20 years ago. They identified 20 tipping points, and crossing each is definitely final. In other words, if we cross these points, the climate will change so much that the Earth's whole natural ecological system will collapse once and for all, and more than 99% of the species living on Earth will irreversibly go extinct. We have to do everything to prevent this!

The tipping point which can be understood most easily is the average temperature of Earth going over the 1.5-2 °C warming threshold already mentioned. With the current 1.1 °C of warming, we are really close to that value. It is especially so if we take into consideration that the speed of changes has been accelerating in the last few decades. (It is true however that the coronavirus slowed it down a little.)

What makes the situation even less pleasant is that by 2019 we had already reached 9 of the 20 tipping points set by scientists (Timothy M. et al 2019). These tipping points are the following:
1. the excessive decrease of ice in the seas of north pole
2. the excessive decrease of ice fields in Greenland
3. the excessive decrease of boreal forests (taiga)
4. the excessive thaw of permafrost
5. the slowing down of the meridional flow of the Atlantic Ocean
6. the excessive destruction of Amazon rain forests
7. the excessive destruction of barrier reefs in warm seas
8. the excessive thaw of the West Antarctic ice field
9. the excessive thaw of the East Antarctic ice field

While reading these lines, many readers may have the instinctive reaction 'There is no big problem—those places are pretty far from us'. Unfortunately, this thinking is incorrect. Climate is global and these changes that seem to be far from us have really serious effects on local climates.

Now comes the hard part: because of the fields indicated by the above tipping points, the escalations of the processes strengthen each other. This is clearly shown by the fact that exceeding the tipping points happened earlier than scientists had predicted. Let's see an example of how the processes affect each other. If there is more ice thaw at the north and south poles, the average albedo of Earth reduces (smaller ice fields reflect back less light) and so the climate continues to warm. Meanwhile, the dissolved fresh water dilutes the water of the oceans, slowing down the circulation currents (e.g., the Gulf Stream). If it continues to slow down, the heating-cooling effect of the ocean will be reduced, due to which the weather will become more extreme. In the meantime, because of the destruction of rainforests, the atmospheric processes change and the rainforests become savannas. These areas can therefore bind even less CO_2 and the warming continues to accelerate. And in turn, these processes continue to speed up the melting of ice caps and slow down the circulation processes in the ocean. Even the destruction of the taiga forests boosts the changes in the system, resulting in even less capacity to bind CO_2, which continues to speed up global warming. Finally, these processes are joined by the excessive thaw of permafrost, due to which a lot of methane, now fixed in the frozen ground, will be released into the atmosphere, leading to an extreme acceleration of the process. These mutual interactions are so powerful that they can cause even an 8 °C warming in the average temperature. For us, anything beyond the 2 °C threshold really doesn't matter, so just imagine what it could mean overall.

In case you don't believe me, then watch UN Ambassador Leonardo DiCaprio's or David Attenborough's and Al Gore's authentic movies about it. At the end of this book you can find a list of movies concerned with climate protection; I can highly recommend them if you long for spending some quality time. And I would like to kindly ask you not to listen to the climate change skeptics or the newly popular "climate realists", who are unfortunately no better than the flat earth believers. They are few in number, but express their incorrect thoughts more and more loudly.

In order to avoid the catastrophe, we have to make the global economic system carbon neutral by 2050 at the latest. This means we mustn't release more GHG into the atmosphere than what we extract. What is much more important is that we are in the final moment; this can be clearly seen from the fact that we have already reached 9 tipping points and the processes are speeding up before our eyes. In other words, we have experienced only small breezes from the storm that is raging nearby. We have no more time! We cannot point at others, say the scientists will solve it, wait for others to take action before taking action ourselves—urgent actions are needed by everyone right now!

Everybody has to do what they can in their lives in order to avoid having their children live in poverty and suffering.

1.2.3. The economic reason

As I already mentioned, the world economy is based on the science of so-called classic economics. I do not say that this economic system is wrong, as from the deepest points of the Middle Ages we managed to get here with only the help of a very simple and effective system. Classic economics has therefore been an important crutch in the dynamic development of the last two centuries. At the same time, I say that this system on our level of development is ancient. For the development of today's society it is a barrier and not a driving force.

One of the basic theories of classic economics is the law of supply and demand. This law says that with the increase of prices demand is reduced, while with the increase of prices, production increases as well. Optimum production is set at the intersection point of these two opposing processes, where supply and demand are in balance. This is shown by the 'i' point in Figure 2. The way we could solve the problem of climate change while keeping to the basic theories of economics is to add the damage and restoration costs of climate change to product prices. This would result in price increases and production would fall, as shown by point 'b' in the Figure. The falling production would cause unemployment and economic recession. So, without modifying current economics we cannot solve climate change, especially since adding these costs to product prices would mean a dramatic, immediate rise in prices. At the same time, social expenses are rising more rapidly from year to year because of the increasing climate issue. The rise of product prices would consequently suffocate the whole system. This is why economists do not take any rational actions against climate change; they are limited by the incomplete system of the basic principles of their science. This is the reason why the whole of 'modern' western civilization takes minimal actions for show without obtaining real results. The restraints of our economic system inhibit our society's ability to change.

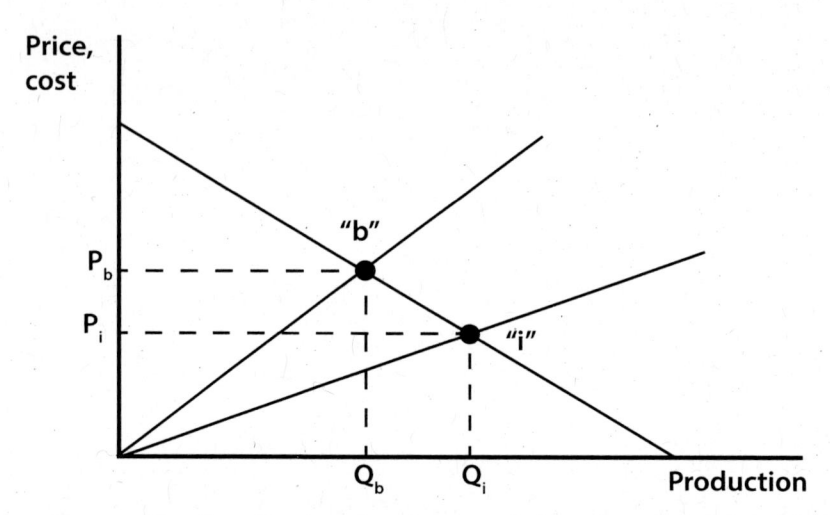

Figure 2: The effect of environmental costs on the law of supply and demand (Zsolnai 2001)

The other problem with the current economic model is toxic competition. According to this model, economic competition is the driving force of development since it boosts efficiency, and as a result social well-being is also improved. This theory was actually correct about two decades ago, but today fierce competition has become a tool of our destruction.

In fact, the entire capitalistic society and our whole economic system are based on competition. All of society is about which company can make the biggest profits and reach the highest economic security by it. The situation is the same among cities and settlements; they compete with each other over who can develop faster and provide better living conditions for their inhabitants. The same happens among countries. In the economic competition between countries, everybody wants to reach the highest standard of living, and the countries with the highest standard are not ready to lose their leading positions. And it happens on the individual level as well: if others in my surroundings live in better conditions, I will feel envy. To reach their level I will force myself into competition instinctively and accept that I will fight, work and push in order to reach the same or higher level than my neighbor. This competitiveness is the driving force behind capitalist society.

Competition has resulted in an accelerating human race on Earth that always wants more and becomes more and more insatiable, and this insatiability, wanting and competition are causing the total and final extinction of natural resources. This is the generator and driving force behind climate change, and what is worse is that it is the same reason behind the bulk of human frustration and negative feelings due to which society's state of mind is so incredibly bad and low. Please think it over: how much pressure is on you all the time because of the expectations of our overexcited society? The unprecedented stress-load of today's western society comes from it. Competition saturates everything on every level and is destroying us continuously without our being aware of it.

In my opinion, the current economic system is not capable of solving climate change or attaining world peace. This stuffy system can be mended and patched up, but it will still be a worn-out, shabby fabric. It's high time to gradually put our economy on a new footing, the way to which is going to be introduced in later chapters.

1.2.4. The biological, ecological reason

About 7.9 billion people live on Earth at the moment. According to the different population growth models, this number will peak at 9 to14 billion by 2050. There have been several studies in the last decades dealing with the price we will have to pay to supply food for so many people. The researchers all had almost the same results: in 2050, 80% of the terrestrial ecosystem would have to be used for food production to feed so many people according to current eating habits (David Attenborough, 2019). We have already destroyed 70% of Earth's natural areas, but only a part of this amount was destroyed for food production. If we have to destroy 80% of it because of food production by 2050, and we have already destroyed plenty of areas for other reasons, what will remain for Nature? How long do you think Humankind would go on without Nature? We are running to the end of a one-way street faster and faster, aren't we?

To show what this all means, there is Einstein's famous viewpoint that after the extinction of bees, the human race would live for a maximum of four years (Marinov 2013). The condition of our ecological systems and climate change have resulted in a reduction of bee populations that has never been experienced before (European Parliament 2008). According to some predictions, bees will be extinct between 2040 and 2050 if we continue this way. This single fact is enough to finally believe that we are in extremely big trouble. How serious is it? In China, mini pollinating drones are being developed, and thousands of people perform pollination by hand in the spring instead of bees.

1.2.5. Habitat and vegetation fragmentation, soil degradation, biodiversity reduction

More and more literature and movies are dealing with the soil as a key point in the fight against climate change. The soil degradation caused by humans in the last century may even be responsible for 30% of the redundant CO_2 in the atmosphere. (Paul Hawken 2020)

In a handful of good quality soil with a healthy structure, there are more organisms than the population of Earth. Soil is a huge habitat. In 100 m^2 of natural soil there can be 1 ton of living biomass. Unfortunately, the quantity of chemicals released by intense agriculture, the sowing and soil erosion caused by it, monocultural production and several other causes lead to the destruction of soil. According to estimations, at this rate all the soil will become infertile in 30-50 years. **In other words, we have only 30 to 50 sowings and harvests left.** There could be serious famines even within our lifetime if we continue in this way. Please, combine the information written here with the information in the previous chapter. Of course, this problem can also be mitigated with a change to regenerative agricultural methods. That is why it is crucial to change the way we will live in the future compared to how we have been living so far.

A more serious consequence of soil degradation is that with the reduction of biomass present in soil, a lot of CO_2 is released into the atmosphere (Paul Hawken 2020). As if it is not enough that the amount of food that can be produced will fall dramatically, global warming will also speed up.

The other very serious consequence of soil degradation is the reduction of biodiversity not only below the surface, but also within the ecosystems living on the surface. David Attenborough (2019) also makes it clear that **the bigger the ratio between biodiversity on Earth and the areas covered with Nature is, the more CO_2 leaves the atmosphere.** So the reduction of biodiversity is not only harmful because plenty of species which had been developing through millions of years will disappear forever from Earth, and not only because what is produced by several species could be remedies for some future human diseases; it is also harmful for the reason that the reduction of biodiversity causes the dramatic growth of CO_2 emission into the atmosphere, which speeds up global warming. The process is autocatalytic. The more natural areas and soil we destroy, the less CO_2-binding capacity the Earth will have, and the poor monocultures that replace the destroyed wildlife will bind only a fraction of the CO_2.

The habitat and vegetation fragmentation just makes the problem bigger. Just look at a satellite image of North America or Europe on Google Maps. The world looks like a chess table. We have cut up the Earth with straight boundary lines. Everything became possessed by people and we believed that we could do anything we want on our properties. Unfortunately, we were wrong!

The result of our activity is that the remaining natural areas have become like small, isolated islands among the huge cultivated areas. This phenomenon is called habitat and vegetation fragmentation, according to which the natural areas are divided to separated parts. The habitat and vegetation fragmentation itself is enough to reduce biodiversity in these areas; this is true even if these areas are so protected that there is no human activity on them. The reason is that every species has a minimum need for its own living space. If the size of the untouched natural area does not reach the size of this needed living space, the species will move on or die out even without any harmful human activity. And as we know well from David Attenborough, the reduction of biodiversity speeds up global warming—another autocatalytic item in the system.

The situation is that there are a lot of solutions. We have plenty of opportunities to do something about it. But if you are unsure whether a human activity is climate friendly or not, you will get the answer due to this common rule: **Every human activity that increases biodiversity, improves the condition of the soil or reduces habitat and vegetation fragmentation is climate-friendly. (The opposite of this statement is also true!)**

1.3. Vision - or the realistic future according to my opinion

The previous part of this book introduced what the world would be like if we did not change. And I also tried to describe that we are deep in the processes which we cannot stop now, but we have to mitigate their effects. We cannot wait any longer; we have to take action right now. The following part of the book is going to focus on the solutions. However, a boat that doesn't know where it's going has little chance of reaching its destination. It is not helpful if we want to flee from the alarming effects of climate change but we do not know where we are heading. Thus the boat needs a destination and its captain needs a compass. The compass is this book. The destination is the future image that I know human beings could and would reach!

Before I introduce you my vision of the future, it is highly important to know that we are totally prepared both scientifically and technically for the future that I am going to disclose to you. It is so, even if at first my thoughts seem to be utopian or naive. I am also going to tell you how we would get there, but let's first look at the right target.

We are in 2151. The Earth has been cooling again for some decades. Humankind realized the situation at the last moment, in around 2020, and started making dramatic changes. As a result, society and the economy went through huge changes, not to mention the common level of people's happiness. The fierce competition is now over. There is no money or private property in society. Please do not be afraid: communism is not going to return! People live according to a long-sustainable social system based on basic human values. The society that was based on personal property and wealth was changed to one based on motivation and service. Every person works only when they want and as much as they can to do others enough good.

Everybody lives in the greatest comfort and at the highest level of technology. All the basic needs of every person are provided. As there is no competition where people squash each other, people instead live according to a realistic value system and realistic wishes. It is not necessary to strive for redundant power and excessive security. The base of their happiness is to be useful to others and to Nature. The reasons why they work is that they like their work, they like to be useful parts of society, and they like to live for others. People live in so-called agglomerations. Agglomerations are surrounded by so-called natural areas. Each agglomeration is in balance with the Nature surrounding it. Nature has its own property. Every natural area is the property of Nature. Nature also has rights which are valid in the natural areas. Nature's rights are protected by a global legal representative organization. Humans cannot perform any activities in the natural areas except for research, education and revitalization[9]. All of the industrial, agricultural and other social activities take place within the agglomerations. Between the agglomerations there is just one large and high-performance transport and utility route, which goes primarily in tunnels, to provide the maximum connection between natural areas and to disturb them less. The old-fashioned road networks, which interwove everything, were eradicated within the natural areas in the previous century. At the same time, the presence of agglomerations does not limit the flow of human freedom, traveling or products.

Humanity chose the approach in which spiritual development is its most important task. As a result of this approach and the end of competition for money, people live at a significantly higher level of happiness than ever before. The agglomerations, living happily within

---

[9] revitalization means to support the regeneration of Nature

themselves and with the others, help each other in development and in reaching their mutual, global target, which is the total balance between Nature and human society and to reach the Global Happiness Index. Scientific development moved into an extremely fast pace and humans started to explore space. The first project to colonize a foreign planet is about to kick off.

This vision for the future seems utopian, doesn't it? I know you think that the author is out of his mind because this vision is impossible. But I have good news for you: it's quite easily reachable! How? By the end of the book you will see the big picture! But before we start on this common journey I will ask you a question: wouldn't it be better to live in a world like the one I've described than in the one we now have?

1.4. The overwhelming Adam Smith vision - or how got so far

Very serious parallels can be found between our current situation and its beginning phase. This is why it is crucial to see how we could get into this trouble where we are now.

In the times after the Industrial Revolution, the realization of Humankind's giant Nature-altering power took place quite soon. During those times we were being fed spiritually by an ancient fear that was based on the fear of Nature. People did not want to live together with Nature; they wanted to be separated as much as they could and wanted to create living spaces where their comfort depended as little as possible on the changes of Nature. In western civilization the idea of 'happiness' did not mean living with Nature in the most harmonious way, but meant the desire for a way of living as independent from Nature as possible. However, even though several hundreds of years have gone by since the Industrial Revolution, the approach to Life of the western part of Humankind has not changed at all; indeed, it has infected the whole world up to this day. Almost every person on Earth follows the same idea. Meanwhile, our strength and power have grown to the national and then to the continental extent, and have become globalized while growing on. The nature-destroying impacts of Humankind have reached a global level. Just think of the problem of the ozone hole, global soil degradation, the unprecedented reduction of biodiversity or global climate change. In parallel, our society has become globalized as well—just think of the internet or our global commercial-financial systems. In spite of these, our way of thinking, social-economic systems and our fear of Nature work according to our old habits. It is high time to change! It is high time to give new directions to Humankind's development! Unfortunately, our way of thinking has not changed fast enough compared to the speed of economic development.

I am convinced that the problem of climate change can be solved and that Earth can be healed. The aim of this book is to provide an obvious and sure solution of how to sort out all of humanity's environmental problems. At the same time, we will see that with the help of this guide we can get closer to both individual and collective happiness. If the development of our social system can take a new direction and Humankind follows the keynotes given by this book, people will be happier in a much higher proportion on the planet called Earth. We will also recognize by the end of the book that being happy is the most environment-friendly thing on Earth. Would not a world be wonderful where People live happily in peace with themselves and others in harmony with Nature? However utopian this target sounds, I have to say again and again that we are already able to achieve it, both technically and scientifically! In this way, this target is absolutely no longer utopian! To give you a sense, I am going to show you a counterexample of Adam Smith, who developed the basics of today's economics in the second half of the 1700s and started the incredible economic development of the last 250 years.

Based on the logic of the principles of Newtonian physics, Adam Smith created the base unit of economics—the endlessly hedonistic man—and related to it the theory of limitless growth. He built up the main parts of today's economics on it. The scientists and philosophers at the time heavily criticized the Adam Smith model. They stated that it was not suitable to describe economic relations since man could never be endlessly hedonistic. They believed that because people live within a society, the social, religious and family expectations and the community's ethical norms would always stand before hedonism for the individual. This was what made us human. Average people were exactly like that in those times; they did not think about the natural limits of limitless growth because the power of Nature seemed infinite. By the way, hedonism was considered to be one of the most disgusting features a person could have.

Today, selfishness has reached incredible levels. The Adam Smith model grounded a modern economic system that was based on the maximum enhancement of human selfishness. That is why this model—and with it today's economics—are called the mechanism of selfishness. As a result, after 250 years the endlessly hedonistic man is not an unbelievable phenomenon but has become the ideal. Just think about: to what extent are those executives, top brokers, politicians, and actors whom the greater part of the population look up to and consider as ideal hedonistic? The word endlessly approaches reality, doesn't it? The Adam Smith model seemed to be utopian and impossible 250 years ago, but now it has come true!

In these days it is literally desirable and cool to be a hedonist, while a hundred years ago it was disgusting and meant to be one of the worst insults. The fast spreading and success of the model was caused by its simplicity and easy-to-use feature for the then-developing economies or industries going through a big change. The economic model became widespread, eventually becoming the economic system of the world. The problem really started when this fairly narrow and simple model developed to be a crucial part of our culture. It resulted in the current consumer society where the individual fell to become a simple consumer and workforce component. It is so true that most of the time in our over-rationalized world, the individual does not think more of themselves than that. **We consume because we think it gives us happiness, and we work more to be able to consume. However, human beings are on a much higher level—a much more colorful and wonderful entity than to underestimate themselves like that.**

If we think about the process of how such an economic model managed to become the world's cultural-social system and how it modified people's approach to life, then why couldn't a current model gradually give a better and proper direction to Humankind's development? The Adam Smith social-economic model was at the time said to have a philosophical and utopian explanation, however today a whole scientific discipline is built on it, which is taught at the Faculties of Economics of many universities. It is not a secret that I want to highlight that if the over-rationalized human race socialized in the current average consumer society reads this book, they will find it utopian—as through their filter the world has nothing to do with it. That is why I ask for your openness while reading this book. It is not important what we are like now; the important thing is what we want to be like! The first step of every change is to imagine what we would like to be like. The Adam Smith model was at least as different from the mainstream thinking those days as mine is now. In spite of this, the thought can be good and the direction correct!

The beauty of my system is that it is simple, and the gradual transition from the current system to the new one is adaptable and solvable on every level of the social and economic systems. From this aspect it resembles the Adam Smith model, but it is better from the viewpoint that it fits more carefully to the basic features of human soul and to Life's rules.

If while reading this book you have the feeling that the gap between my future vision and our present is unrealistically big, please think about the fact that it is not the person of today who has to change the world entirely. Instead, it is the person of today who has to set Humankind's development in the right direction **RIGHT NOW**, so it is up to us to start the change. For this reason, we have to understand what 'medium-term future' means, which will show the right direction for Humankind. The parallel between the Adam Smith model and the one described by this book is that both of them were written on the brink of a big change, and both of them imagined a brand new systems approach almost totally free from the actual stuffy and harmful habits.

The targets written in the book can be reached within several generations. In fact, due to our global information systems, such a change could happen faster than in the time of Adam Smith and later. Moreover, it can have its positive effects in a really short time, so the model will have a positive impact on people's happiness today.

I hasten to add that I am not at all anti-development—I see myself as a development supporter and believer! This will be clearly seen from the detailed description in this solution package. What's more, I believe that the development that has taken place since the Industrial Revolution was a really important and necessary part of human development. Without it, we could not have reached the level of development both scientifically and technically from where we are able to move on to the next level of development. But the situation today urgently requires that we set the new directions of change. The old dogmas and habits have to be replaced by new directions of development to enable Humankind to continue its development and not to wipe itself off from evolution and the history of life with its selfishness and narrow mindset.

1.5. The dividing of our tasks between generations

It will certainly take more than a decade for Humankind to reach the dream which may seem utopian now, namely that individuals will live in peace and harmony with themselves, with others and with their environment. In my opinion, it will be 3 to 5 generations before this dream becomes an available, physical reality. Our responsibility is to choose the right directions for our development and to encourage subsequent generations to continue them. Obviously, the development of the system draws on the correction and further development of the bases written down in this book, and the principles mentioned here need to be adapted to the different levels and sub-systems of society, giving us a lot of tasks in the future. It is just a guide that shows the right main directions and describes them in a simple, everyday way.

As I mentioned, there is an entire economics science based on the fundamentals of Adam Smith's economics, so why couldn't a separate science be built in the future on such current concepts? I have called this discipline 'the science of equilibrium' showing the point of the messages of this book.

Our generation's task is to save time! Our task is to postpone the start date of global collapse which is estimated to happen around 2040 or 2050 by 20 to 30 years. This is our generation's task. With it, our generation provides time and a chance for the following one or two generations to stop the deterioration-destruction-climate change processes. The task of the two generations coming afterwards will be to reverse the trends. That way, in about five generations from now we will be in balance with Nature and will live in a society that is much happier and much more balanced than now.

CHAPTER 2

Patterns of thought needed for a solution

In this chapter I am going to show you several novel patterns of thought which are necessary for seeing the real solution system as a big, clear picture. It is important to know that everything you will read in this chapter has been scientifically proven. So please keep this in mind if you find something too spiritual, weird or silly at first, and please stay open! Just because something is new or different from what you'd thought doesn't mean that it's bad. One of the first steps to attaining climate change is human openness; it is capable of setting us on new, proper paths. An important message from the previous chapter is that the problems **of climate change and the unrest in the world are caused by the wrongdoing of our society's widespread patterns of thought.** In order to be able to change **we have to change our mindset first!**

2.1. What is the second step?

If we start a chess game with the wrong moves, our chances of winning at the end are reduced heavily. We keep on reducing our chances of winning with every wrong move. The fight against climate change is like a chess game. However, Humankind's game has a huge difference from a chess game: in this case the Earth will be the winner regardless. In one case, Earth wins and the human race loses; in the other case Earth and humans win together in a friendly, smiling draw. I am personally working for the latter game result every day, and this book was written for that reason. Earth has survived all the big extinction processes so far, and a new ecosystem has always developed from the levels of lower forms of Life. Earth doesn't really care what happens to us, but it matters to us! It is high time to step actively together for our common future. **Life has only one task: to maintain life.** Whoever turns against life will therefore die out. It's that simple.

The first move in our chess game has not succeeded very well. Our first and most important task was to make humans aware of the fact of climate change. As long as people did not believe that climate change was real and that we are responsible for it, scientists were warning us in vain. We have finally reached this point. But if we've reached it, why did we make so many wrong moves in this game? The answer is really simple: it took 50 years for about 90% of Earth's population to believe that this problem is real and very serious. This ratio was 93% in the EU in 2019 (Conscious Buyer 2019). The remaining 10% belongs to the climate change deniers and the newly popular 'climate realists'. The latter used to be climate change deniers, but they no longer dare to say that climate change is not a fact. They are therefore using some convoluted reasons to claim that climate change is not a big problem. It is important not to listen these ideas, but just to look ahead and do our job: to do the right thing and fight against climate change.

If we had succeeded within 10 years instead of 50 to reduce the numbers of climate change deniers and climate realists to their current low level, the first move of the chess game would have been right, as we would not be in this situation. Earth answered with continued warming, and heatwaves, hurricanes, droughts, forest fires and so on became more frequent. When I see the increasingly dramatic weather events in the media, I have the feeling that Earth is fed up with Humankind's harmfulness and wants to wipe us off its surface. But obviously, Earth does not have a consciousness[1] so it does not do these things deliberately. But Earth's ecological and climate system have already started to work against us. We kept the Earth off balance and now it is searching for new balance where there is no longer a place for people. **Earth's climate system pours more serious warnings on us every day,**

---

[1] However it is truth only according to the realistic view

the messages of which mean only one thing: **wake up, humans, or this is the end!!!**

As we made the first move in our chess game in a bad way, the second move needs to be done properly or the chance of winning shrinks dramatically. What's the second move? **Taking responsibility based on being well informed.** Now that we at last know that climate change is a fact and all but a few people agree on this, we have to understand we cannot ignore our responsibility because **7.9 billion people are responsible for climate change.**

It's my experience that people are behaving the same way towards climate change as in other situations. If people are hurt by something that they don't like or is inconvenient for them, what is their ego's instinctive reaction? Passing on responsibility and understating. People in these situations instinctively search for some excuses, which means accusing others or acting as if they aren't interested. Taking responsibility for our deeds happens only on a higher level of the soul when the hurt ego calms down inside and we start to think that our initial reaction may not be right.

I can see that Humankind is at the same stage regarding climate change problems. Most people think: 'All right, I accept that there is climate change and that we people are responsible for it. But it's the task of politicians and scientists to do something about it.' Of course, some people accuse global companies, some the oil companies, and some think they are too small to do anything. Unfortunately, we cannot pass off the responsibility to global companies, airlines and politicians. On one hand, we established and own these companies. We elect the politicians. On the other hand, a lot of the gas that causes climate change is released into the atmosphere as a result of our own decisions. For instance, if you choose between a chicken and a beef burger, or between flying and taking a high-speed train, you can save more than 95% in CO emissions. The difference is the same if you buy a local or a Canadian wine in a Hungarian supermarket. In a lot of situations, we can make our decisions in a way that can directly save the Earth from a huge amount of GHG released into the air.

In order to decide in the right way, two factors are needed: the first is to be well informed, and the other is taking responsibility.

Being well informed is necessary because if we are not aware of the consequences of our deeds, we will perform them without thinking about whether we're doing the right thing. For example, until I knew that every time I eat chicken burger instead of a beef burger I save climate gas emissions equivalent to 25 km of driving, I ate beef burgers without any bad feelings. But after learning this, I always chose chicken burgers or fish burgers whenever I visited a fast food restaurant. Today I am a vegetarian, so now I do much more against climate change every day. From the very moment **we become aware of the consequences of our deeds, we have the chance to decide in a proper way.** Most people are not aware of these facts, so they do not even have the faintest idea that they make bad decisions. However, I am convinced that people basically have good will (of course there are exceptions). In other words, **if people are aware of which choice is right, most of them will make the right choice.**

Taking responsibility is important from the point of view that if I know that I am making the wrong decision regarding climate protection, I will take responsibility and can still have the option to balance my scales. Here's an example. I am aware of the fact that if I travel by plane, I will generate a huge surplus of GHG emissions compared to train travel, but I still do it. The reason could be that I am in a hurry and don't have time to travel by train. In this case, I can still compensate for my wrong choice by supporting an organization set up for this reason that plants 50 trees. It was important to me to reach my destination fast, so I made the wrong decision in terms of climate change. But as I take responsibility for my deeds, I will fix it with other deeds. (By the way, this creative idea is from an enthusiastic reader of my blog.) Of course, it can happen that somebody is totally uninterested in this and will choose flying and beef burgers, and isn't ready to deal with this whole thing. Obviously, there will always be people like that, but I know that most people are not.

So the next move in our chess game is to read and speak even more about what kinds of climate change consequences our deeds have.

It's important to inform each other through more and more channels. You can find a continuously widening range of information regarding the topic in the blog and website mentioned at the beginning of this book. If we have already recognized the reality of something, we should make our decisions with awareness of our responsibility day by day. If we still choose an irresponsible solution, we should find a way to fix it! Please keep in mind that you can be climate conscious only through gradual self-development!

Why is this the next move? Because **this way can we reduce our GHG emissions to the greatest extent** and the most efficiently without changing our standard of living dramatically. By eating chicken burgers instead of beef burgers, I did not become less happy at all. Billions of people's individual decisions open up huge opportunities. **Wastefulness and not taking care of the consequences of our deeds are the two most powerful kinds of environment-destroying power.**

So from high-level politics to the individual level, our most urgent and **important task is to inform each other,** and scientists should develop a lot of such decision-making alternatives. That is to say, the average person understands these simple messages most, as they have neither time nor energy to think further about scientific results and project them onto their lives. By the way, I am also working on a book which contains such examples, and I hope I will be honored by your attention then too! I also recommend with all my heart *Paul Hawken: Drawdown*, in which you can find a hundred kinds of practical opportunities for the fight against climate change.

2.2. The mistaken paradigm of competition

I heard on the internet from a Buddhist philosopher that competition can only be evil. At first I found this thought to be really weird, as our entire economic-social system is steeped with competition. But with time it matured in my conviction and made me realize that he was absolutely right. In a competition there is always a winner and one or more losers. We usually make the mistake at the societal level that during a competition we always pay attention only to the winners. If we watch a football match, everybody celebrates the winners afterwards, and only a few care about the losers. It's the same within economic competition or if we set an individual goal; we always imagine ourselves coming out in first place. There are always losers as a result of competing with each other, and the experience of losing is a dramatically tough negative energy package. This negative experience is always at the core of developing shame, guilt, apathy, fear, jealousy, desire, anxiety, anger or rage from losing or defeat. Failure to beat one's competitor usually develops a mixture of these feelings. With time, one of these feelings becomes dominant and creates the world of our feelings. From that point on, you do not live in the here and now, but in your reactive and subjective reality. Just think about how many times you felt guilt, shame, anger, or any of these feelings because you came out as a loser in a quarrel or any other competitive situation. By contrast, winners are always proud of themselves for coming out positively in a situation; they feel that they are better and more than their competitors. As a result of this pride, the ego is strengthened and selfishness grows with it. It means that you can never come out of a competition in a positive, Life-supportive way. Instead of competition, the right way is a cooperation that originates from balance, peace, joy, harmony and adaptation. There should be no competition in the future because as long as there is competition, negative and soul-degrading energies will always arise.

At the social level, balance is expressed through cooperation—through cooperating activities carried out together for each other's goals. The paradigm of cooperation negates the validity of the Darwinian model, which is based on competition, even at the evolutionary level! Darwin is totally mistaken in his theory (according to Dr. Bruce Lipton, the father of epigenetics): life is basically meant to be cooperative, not competitive. It is not competition, but rather the willingness to engage in cooperation and its level and extent that drive development and change forward. Current economic competition creates a system that destroys Life, as it is continuously developing billions of negative energy packages in people. We can actually have a strong economy without competition; furthermore, our society can also be strong without it.

You may ask what would happen with sports if there wouldn't be any competing in future society? There would be sports in the future, but no competitive sports. Exercise would be done for the sake of the pleasure of moving, developing our bodies' self-control, and as a tool for spiritual development. Sport is wonderful in itself, but it is also destroyed by high-performance sport which is forced into extreme competition. When we play basketball in the street happily, then sport is a wonderful thing and a wonderful experience of being. But the moment we start playing on a professional basketball team and find ourselves in a crazy, competitive situation we tear ourselves apart, the positive experience of *being* ceases, and from that moment only the negative energies coming from competition can be experienced, including the craving to win. It destroys Life at the individual level. It's no accident that high performance athletes usually burn out by about the age of 30 and cannot continue.

If we examine a match from the fan's perspective, we can come to similar conclusions. The fans jeer at the other team and hate the other team's supporters. Competition generates Life-destroying energies here as well. The happiness of the winning team's fans is a short-lived one, which strengthens the ego in a latent way, and also develops Life-destroying energies with it.

Competition will disappear from society gradually of course—not from one day to the next. The person of today would not be able to exist without competition. The first sign of change will be the moment when everybody supports their own team without booing the other team or the referees. It will be such a social level of the development of the human soul, where the winner attitude of competition is dominant, and degrading each other is no longer an aim. The same will happen in economics and in other segments of society. The first temporary step will be when competition is gaining some culture. If nobody attacked their competitors and wanted to seem better at the expense of others, this alone would result in a much nicer society than today's.

2.3. Every Life is equal

Competition is the manifestation of showing ourselves to be more and better than others. It is one of the biggest problems of our society, which is enhanced and overheated by competition, but unfortunately it is the basic attitude of people today. The continuous strengthening of the ego enhances the feeling of separateness in people, due to which we always want to be superior to others—and this is what leads to unhappiness! As I am going to prove the following in detail later on, **unhappiness is one of the most climate-destroying things in the world. The most important basis for happiness is in close human relationships.** We make ourselves unhappy by alienating ourselves from other people by highlighting our differences.

This book is going to proclaim the theory of equality throughout: every person is equal—indeed, every living is equal! A blade of grass has the same right to have a happy life as you and I have. Our right to happiness is independent of our color, religion, what we look like, or if we are heterosexual or LGBTQ.

In future society, every person will respect diversity. Today most people judge anybody who thinks differently, believes in different things, or looks different. This is the disrespect for diversity that steeps our whole society—that causes so many tensions, wars and an incredible amount of negative energy. Just think about how many wars have been caused because of oppositions coming from nationalities or religions. Every manifestation of negative energies is destructive, so every thought or statement that disrespects diversity is destroying Life. And the destruction of Life accelerates climate change. **One of the most efficient tools of climate protection and reaching world peace is respecting and accepting diversity! Live your life according to the belief that every living thing is equal** and you will have already started to live a Life-supportive, climate-protecting, and more peaceful way of life!

2.4. The new level of evolution

I was talking to an ecologist—highly respected by me—about the problems of climate change and overpopulation. His opinion and thoughts are really clear, logical and obvious.

During the roughly 4-billion-year-long evolution process, millions of species developed and then became extinct. Several factors could have caused the extinction, but in most cases there were two reasons. One of them was a change in environmental circumstances (warming, desertification, ice age, meteor strike etc.). The other reason was the phenomenon of overpopulation. As a matter of fact, **every species will breed as long as their environment can sustain it.** If a species with a high population has reached the limits of its environment, the result would be the dramatic reduction of the species, often lasting until its total extinction. Life went on, but the species disappeared from Earth forever, and other species came instead. What is the situation today? Man is overpopulated, and has reached—even exceeded—the limits of the Earth. Furthermore, the environmental circumstances are changing dramatically; global warming is accelerating. From the viewpoint of an ecologist the result is obvious: a sudden decrease in the species, then extinction. The ecologist mentioned above is absolutely sure that we are going to die out, as it follows logically from the basic laws of ecology.

I partly agree with the ecologist expert as, unfortunately, it is a very realistic outcome. If we people do not overcome our basic evolutionary limits, we really will disappear from Earth. Then life will gradually recover on Earth without humans, but everything will go on. Our short two hundred thousand years within the 4 billion years of Life on Earth will have been just a moment. So Life does not really bother itself about us... the existence of one species or another does not affect Life.

The evolutionary limits of Humankind means that we live without any consideration; we increase our population and consumption unconsciously until the point when everything suddenly collapses and the process of Humankind's extinction starts irreversibly. In the time of scarcity everybody will become more selfish, which makes the process even faster.In other words, the same thing will happen to us that has happened to many millions of species so far in Earth's history.

But I am convinced it will not happen. Man could be the first species in the history of Life on Earth, as it is known by us, which could rewrite the basic evolutionary rules, as proven scientifically. Humans are the first species possessing strong consciousness, which we are able to control. I'm not referring to our ability to think. It may seem surprising at first, but I'm talking about our spiritual consciousness. But please don't be afraid! A lot of rational people would put down this book at this moment. I promise I am providing only scientifically proven facts!

What does spiritual consciousness mean and why is it the key to a solution? Spiritual consciousness means living one's life consciously and working toward spiritual purity and spiritual development consciously every day. A conscious person is characterized by **instinctive moderation**, which is one of its keys. If you take a child to a toy store and ask them what they want to take home (if they can choose anything), it's most probable that they will want half the store. However, an adult sensibly considers that such spoiling would do the child no good and is also aware of the financial consequences of such shopping. The child is longing instinctively for the many shiny, exciting toys while the adult decides consciously, and only one or two toys land in the shopping cart.

Humankind has been behaving in the last more than one hundred years as children in a toy store. We have wanted everything because we believed we could take anything we wanted from the shelves of Earth. Humankind has become as selfish as little children—maybe even worse. While this is natural for a little child, it is not so for adults. Our 'parent' is the Earth, which sends us warnings all the time. In spite of this, we have been behaving like little children. We have not understood why we can't have whatever we want—**RIGHT NOW!**

A natural change in our behavior is the direct consequence of stepping into the next level in our spiritual consciousness. This increasing level can now be measured and is scientifically proven. In order to introduce this system, I am going to tell you in the following chapters about the connection between climate change and the human soul, and about the measurable levels of vibration of the soul. Afterwards I promise I will return to more practical issues, but this topic is really important for understanding the explanation of practical issues later on.

Those who live on a high level of spiritual consciousness **take responsibility** for their deeds and are aware of the harmful consequences of their way of life and do everything to reduce them. The individual does this instinctively because the conscious person is moderate instinctively if they feel their deed causes more harm to the society than good to their life. The more consciously we live, the less extensive is our **over consumption** and **selfishness.** The more conscious we are, the more we respect ourselves, other people and Nature. What is more important, a higher level of consciousness raises acceptance instinctively, which radiates from the inside out. As a result, we are less likely to want to transform and reshape the world around us. This incredible desire of transforming comes from the lack of balance in Humankind's soul. As we do not accept ourselves, we cannot accept our environment. We instead invest all of our efforts in transforming the world, but unfortunately not in the right way and not in the right direction.

The right way is for us to leave behind our personal materialistic, money-grubbing selfishness that is focused on appearances, and to find the way of spiritual development. Spiritual consciousness is one of the guarantees of Humankind's survival, further development and well-being. It's time we rediscovered what has been known by Eastern philosophies for thousands of years.

Be more conscious. Be happier. Raise your children according to this. Guide evolution in the right direction with it and save the Earth in a form suitable for human Life!

How can the correctness of this statement be proven by figures/numerically? And how is it connected to the vibration level of the soul? I am going to get back to this, but first it is important to understand how an individual's happiness is related to climate change and world peace.

2.5. To what extent are you a Life supporter?

I have already used the terms 'Life-supportive' and 'Life-destroying' in previous chapters. Now it's time to speak about them in detail.

We live in a dying world. The space for Life is be reduced continuously. More and more people do not support Life but destroy it with their thoughts and deeds. If we want to give a picture of our lives from this aspect, then the most effective way is to observe the results of our activities. It is obvious that we have to do a lot of Life-destroying activities in our lives: we have to cut down trees to build a house in their place, we get into our car and with every push of the gas pedal we vent out exhaust emissions into the atmosphere, or simply by buying a car we cause the destruction of dozens of tons of natural resources indirectly. Plenty of examples could be given. Our lives have a Life- supportive side as well. We have children, plant trees in our garden, take part in communal garbage collecting, install solar panels on our houses, etc. If you take these things into consideration, is the result Life-supportive or Life-destructive?

Most probably, most people today belong to the Life-destroying group. This is easily seen, as the natural resources and ecological systems are decaying dramatically, and biodiversity is reducing rapidly. If more people were basically Life-supportive, this tendency would change. One of the most crucial basic principles of the fight against climate change and for world peace is that **we have to become Life-supportive.**

We are living beings, so it is obvious that the result of our lives must be Life-supportive! We are also part of Life. If we destroy Life, we are 'traitors' and join the 'enemy': the Lifeless world. Even though the living and the lifeless world live in cooperating symbiosis with each other, we must not support the predominance of a lifeless world or we upset the balance. This idea is supported by the fact that those who live in a Life-supportive way are much happier than those who are on the other side.

The biggest driving force behind destroying Life is selfishness. Dear reader, I know that you too have had the thought that we have to be selfish in the world. You are absolutely right! But you can decide about the extent of that selfishness. You could strive for the minimum needed level of selfishness, or for its maximum level. Unfortunately, today's trends are about maximizing our selfishness, and there is a corresponding increase in Humankind's anti-Life feelings and the destruction of Life. It's not that we don't need to be selfish in some situations. In a situation where our physical integrity is at stake, the right way is to activate our selfishness (self-defense) instinctively, and this is the basic code from an evolutionary aspect. However, plenty of manifestations of selfishness have become massive in the world, like the desire for money, lust for power, a lustful way of living and many others. In fact, these are addictions that have been developed through the distortion of human society.[2] The power of the ego has reached a level where the individual is less interested in moral, ethical or religious rules. Other people or the environment are less and less important for the individual, who focuses only on their own self-interest and wades through anybody or anything to reach their goals. This is the basic driving force of the Life-destroying power, namely the infinite desire for self-satisfaction or self-interest. The problem basically is that the limitless desire of the ego cannot be satisfied. Whenever we reach or get something the ego 'has a short rest' so the soul seems to become happy for a while. Then the ego soon sets other targets and desires so the happiness is only temporary; in fact it is just a mirage. **The ego always wants more!** If there are about 8 billion insatiable, limitless egos living on Earth, how much chance you think terrestrial Life has to avoid dying out?[3]

    The result is an ever-speeding world where Humankind chases the production and consumption of goods like a maniac. The bigger amount of these goods is not about satisfying real human needs. Many examples could be mentioned. I know people whose closets are as big as an average person's bedroom, and they have hundreds of clothes and shoes which they have only worn once in the fitting room. Shopping has become a form of addiction. We want to possess even more things, and because we're not addressing the healing of the real problems of

---

[2] Addiction means spiritual dependence. You can read about games in appendix 2 and I truly recommend the In the Realm of Hungry Ghosts by Dr. Gábor Máté

[3] You can read more about the ego in Appendix 1.

our souls, this even more greedy wanting tries to fill a bottomless hole. The system has degraded us to be only consumers, which is a more modern form of slavery. We are led through our addictions and meanwhile we believe it is good for us. Of course, this process is unconscious since we only see that possessing and using newer and newer goods and services makes us happy. Due to the fact that it causes only short and superficial pleasure, we need more and more. We always want even more and it is never enough for more than a short time. Those who cannot afford it are blinded by the shallow glamour and are longing for such a life. **Almost everybody is addicted in one way or another to the Holy Grail of shopping or to the associated feelings of comfort and safety. Meanwhile, their life is passing by and they realize only on their deathbed how wrongly they have lived** and how many things they should have done differently. They should have listened to their souls, they should have paid more attention to their loved ones; positive spiritual values and spiritual self-development should have been the driving forces in their lives. It is a shame that billions of people only realize this on their deathbeds, at the last moments. **Meanwhile, they have ruined not only their own lives but their children's future.**

The media make people believe that they can get anything or reach anything if they want it badly enough and fight for it strongly enough. This is strengthened further by the profit-oriented companies with their ads affecting the depth of our psyche, exploiting our natural needs. They make us believe that we desire what they offer. Obviously this is not true, but we chase a lot of mirages in the hope of happiness all through our lives. The truth is that we cannot reach anything because we have our own limits regarding our genetic and psychological abilities. For instance, someone born with a slight build cannot be Mr. Olympia, or someone with an average IQ cannot be a world chess champion. As a result of this controversial social situation, most of the people living today are frustrated and longing for the examples shown by a small portion of the population, which they can never match. In spite of this fact, everybody does everything in their Life to be able join that particular small group. To reach these targets, the individual maximizes its selfishness, often even putting its conscience aside to be able to proceed more efficiently to its desired targets. The result? Even more selfish and frustrated people. The vicious cycle is eating up the natural part of the world more and more intensely and faster, as well as all the cultural and other values, while human selfishness is increasing and unhappiness is growing in parallel. Unfortunately, the biggest trap of this vicious cycle is that we increase our selfishness in order to reach our goals in a more efficient way to try to become happier by it.

But most people just do not realize that with the increasing level of selfishness it is not happiness that is growing in proportion to it, but misery. I was living that way for decades and know well that it doesn't work. Selfishness has become so powerful in the world that **most people live their whole life without realizing that they are standing on the wrong side. They do not even notice that they have become Life-destroying.** It can happen that a thought crosses their minds that perhaps it's not way they should be doing things. But the selfishness inside and the powerful ego banish this thought with many reasons, like: 'Why I shouldn't I do it when everybody else does?' Or: 'My little actions are just drops in the ocean. If I changed, nothing else would change.' Or: 'I'm worth it; I've worked a lot to get it', 'I deserve it; if others can do it, then why shouldn't I?' Thousands of such self-defensive reasons could be listed here, with the help of which we can hide away from our conscience and sweep the facts under the carpet. Why do we do that? For several reasons: first of all, it is simpler; secondly, it is much easier to believe the glamorous propaganda created by the short-term targets of politics and global companies' interests.

The truth is that most people have neither the mood, time, nor energy to read between the lines. They can't be blamed for this of course, as the average person does not want anything else but to live in peace and tranquility, and for that reason accepts the actual worldview trends and believes what the owners of the power put in front of them. This is true today, but it has always been like this throughout history.

Therefore our aim **is to control selfishness, to strengthen the feeling of togetherness, selflessness, community perspective and Life-supportiveness wherever it is possible.** If we go in this direction, it will have a positive effect within decades. If you can accomplish this, your personal happiness will improve and the average happiness level in society will grow as you contribute to it. The level of the destruction of Nature will gradually be reduced as well. **And the best thing is that it does not cost money; you only need yourself and a slight change in mindset.** The inspiration should be the fact that your spiritual development is the only way through which you can be permanently happier!

2.6. The respect for Nature and happiness

People have forgotten to respect Nature! As long as we were scared of Nature, we respected it. But since we have been ruling it (of course this can only be temporary) we do not really care about it. We think Nature is for us and we can do anything with it. However, if we think logically, our dependence on Nature is obvious, but considering it from air-conditioned shopping malls, we are distanced from Nature. We have fewer and fewer direct contacts with it. This distancing has caused us to become disinterested in natural issues.

But if we open a little towards Nature again, people will soon realize how good a feeling it is, say, to plant a tree. The person's conscience will feel how good it does people to take part in such social actions. If we take part in a social garbage collecting, it is not only the cleaned forest that will feel relief but also our soul because of the good feelings generated in us. Humans discover again how pleasant it is to do something together toward a common goal for the sake of Nature.

In our western society, people live at the highest level of Life that has ever been experienced, however Humankind has never been as unhappy and lonely as it is today, except for some shorter war periods when unhappiness was more dense and unbearable. People have forgotten the joy of living together with Nature and the joy of togetherness with other people. C. G. Jung pointed out in many of his writings that the **human soul is happier because of the uplifting effects of the natural environment and community life.** That is why I repeat:

the most crucial fundamental element of a happy Life **lies in quality relationships with other people.** According to the experts, you should have at least 13-17 quality relationships to avoid social illnesses or shortages, and in order to feel you really belong to others and you are not alone.

If we get back to Nature and community, it will boost our personal happiness. Huge economic force is not needed to do this: just personal motivation and some cooperation! From our individual perspective, openness, bravery and trust are also needed. I know from my own personal life that it is worth it.

2.7. The common root of personal unhappiness, social unrest and climate change

For long years I thought about how selfish people are because they all search for their own happiness. Now I see that this is the only right way! **Everybody's most important life-task is to find happiness and help others along this way.** As Neale Donald Walsch wrote in his wonderful book series *Conversations with God*, we are not going to be judged by our achievements in life but by the **impact we have had on others.** According to this measure, the average person today can be judged quite negatively as being critical, thriving at the expense of others and destroying Life.

It can be clearly seen that the root of the problem lies in the fact that the way and direction of searching for happiness is not right in most people's lives. According to the materialistic perspective, we think that more money, a nicer car, more traveling etc. are what make us happy. The things we are able gain cause bigger happiness. Obviously, this is not true! At most it can cause bigger complacency, a greater sense of security, a temporary feeling of enthusiasm, but not more happiness at all. It brings only quantitative change in our Life, not qualitative.

So there is no choice: **our spiritual development is our most important life task.** The greater the spiritual balance, self-acceptance and peace we find in ourselves, the happier we will be. Inner peace creates harmony inside and in our environment. What's more interesting, it also reduces selfishness. Devotion and selflessness make us happy. In sharp contrast, selfishness separates us and makes us greedy and yearning but not happy. I know it from my own experiences; I have spent a lot of time on the wrong track. At the same time, the happier you are (in the good sense of happiness) the less environment-destroying you are. I am going to prove this to you in a later chapter. It is enough in this chapter to understand and accept that the **unhappiness of 7,9 billion people, the growing climate change and the unrest in the world all originate from the same root.**

We hardly invest any energy in healing our spiritual problems, neither on the social nor on the individual level. Just think about how much time, money and energy we spend on making our bodies beautiful. During our whole lives we try to keep a healthy diet, we keep ourselves clean, we exercise, buy nice clothes, go to the hairdresser and beautician, shave, some go to the tanning salon, we use a vast amount of cosmetic products, and some even undergo cosmetic surgery to enhance their beauty. We spend a significant part of our lives shaping and forming our look—and live with the exaggerated frustration connected to it.

A person consists of 95% spirit and only 5% body (these are not exact figures: just values for presenting a ratio). It is scientifically proven today that we can examine a person on an energetic level as well as a physical level. If we invest so much energy in making our bodies beautiful, imagine how much time and energy we need to reach and maintain spiritual balance. **Spiritual development** is the key to our future as it **is the most efficient climate protection as well!** Search for spiritual balance and spiritual peace, and help other people on this path!

In order to support your spiritual development, meditate at least 15 minutes a day. Read self-help books which support your spiritual development. Go to a psychologist, kinesiologist or to groups where people with psychological wounds similar to yours go. I could provide further advice on that topic, but this book is about solving climate change so there is no space for that.

As the average amount of happiness rises globally, the addiction will be reduced, so the redundant shopping and consuming habits will be reduced as well. In this way, environmental emissions will decline and the Earth can breathe with relief again. According to my estimates, if people considered their spiritual development as the core issue of their lives, it **would reduce CO_2 and other harmful gas emissions by 2,030% immediately!** It is not that financial well-being and a nice appearance—which the world today is focused on so greatly— are not important. It is rather that we have forgotten about our souls, conscience, and spiritual balance.

To sum up: you can be happier and save the Earth (too)! But for this, you have to change your way of searching for happiness if you had not done it already. The happier you are, the more realistic your consumption will be, the more moderate the environmental effects of your deeds will be, and the more life-supportive energy you will radiate towards the world—and it will impress others in a way you will not recognize. To show that how it works is scientifically proven, I am going to get back to the topic of spiritual vibration levels.

2.8. Manifestation

When I was younger, I once read from a Buddhist philosopher (unfortunately I do not remember his name) that the ever-increasing amount of garbage is the physical manifestation of Humankind's increasing spiritual problems. I was about 30 then and thought that however the idea was interesting, it was a little exaggerated. Such an obvious cause-and-effect relationship cannot be stated. Fifteen years have gone by since then. Today, I consider this statement differently; I am totally convinced that the Buddhist philosopher was absolutely right. Moreover, I can say with absolute certainty that **all the environmental destruction and environmental pollution are the manifestation of Humankind's spiritual problems.** Now I am going to prove it to you.

The human soul has a feature in that the bigger its physiological wounds are, the more extreme behavior is characteristic of it. In other words, the more uncured spiritual pains are hidden in our soul's deep corners, the more irregularly and variably we behave. Most of the time these mannerisms are considered to be addictions, like excessive consumption of drugs, alcohol, or cigarettes; to be addicted to speed, gambling, perversion, excessive lust for power, perfectionism in our bodies; and many others. There is a common feature in these distortions of human attitudes: dependence. The consequence of our spiritual dependence is that occasionally or continuously we get those unstoppable feelings to experience or gain those things or activities again. The deeper the dependence is, the more sacrifices we are ready to make to reach the target generated by our soul. There are people so deeply sunk in their dependencies that they are literally ready to progress to their goals at any cost. People who identify themselves with such behaviors use many more resources during their lives than balanced people do, as the former live at a higher speed, using more energy and producing more waste because on their way to their targets they do not care about anything else. They see only the target and search for the fastest way. Every aspect that does not drive them towards their goals is considered to be a setback.

Just think about how many natural resources are destroyed in vain by someone who is addicted to eating and weighs 200 kg just because of their addiction. How many tons of food are digested when the consumption has no connection to subsistence? The human race is being choked by its addiction to the joy of eating. It's like a drug addict thinking continuously about the possibility of getting their drug even while they're eating, so they feel every minute is wasted if it's not spent on getting drugs. This way of eating will be fast, and the produced waste (cans, packaging etc.) will end up on the ground. In the food addict's state of mind, it is a waste of time to take the waste to the first waste bin they see, let alone to sort it and put it into recycling bins. Similarly, it's not an issue for someone who is ready to commit break-ins or other criminal acts to get their drug that the damaged door, smashed cabinets, and broken commodities will all become waste, along with the energy used by the victims to restore their property—not to mention the amount of waste produced and the environmental load of producing and shipping the goods needed to be bought again.

The Hungarian language expresses well the state of a person who is free from any addictions; it uses the word 'balanced' for them. A balanced person is much more moderate. As a result of their spiritual balance, they do not need anything in particular to feel good about themselves. Compared to an ever-craving person who suffers from addiction, a balanced person uses much fewer resources and generates much less pollution in the environment if we take their whole lives into consideration. In addition, balanced people are more likely able to live in peace with others, creating less fear, frustration, anger or other spiritual pain in others. As a result, they cause less indirect environmental impact as tensions caused in other people result in new extra environmental impact. These work like waves on the surface of water. A frustrated person has a continuous impact on their environment regardless of whether they are aware of it.

This line of argument so far does not explain why every form of environmental destruction is a manifestation of Humankind's psychological problems; we have only proved that spiritual problems cause further environment destruction. However, there are sadly very few people on the street who are balanced and free from addictions. Almost everybody is yearning for something, so the surplus of environmental impact coming from this reasoning is significant. That is why **turning to our spiritual development at a global level saves a significant amount of GHG emission and raises the level of peace very quickly.** In other words, it is an extremely effective tool. However, an average balanced person's ecological footprint in our western world is even bigger than what the Earth's natural resources can handle without any significant damage. Could it be stated that if every person on Earth were balanced and free of addictions, the excessive pollution of the environment would stop? The answer seems at first to be 'yes', but if we go deeper into the topic, it turns out not to be true. Even if this answer does not seem logical at first, it is the correct one. The other parts of environment destruction should be searched for within our social system and in the improper development of Humankind's mindset. We have been scratching the surface so far, but let's take a deeper look into the problem now.

The fear of Nature became ingrained in us before ancient times. People could not stand being vulnerable to Nature. The increasing development of the frontal cortex in our brains, which is responsible for rational thinking, resulted in more than just the ability to think logically; we started using the main part of our rational thinking to maximize our safety, and unfortunately this remains so today (B. Lotto 2017). From another perspective, the main application of our rational thinking was and is the escape from our fears. The instinct to strive for safety is older than rational thinking as is rooted more deeply, so rational thinking became the tool of the instinct to strive for safety. It is a natural evolutionary strategy on the individual level. Think about it: your brain is almost continuously scanning the alternatives of the future and tries to navigate you in the right direction to protect you from all the things that are not beneficial to you. This cognitive mechanism and the caveman's constant fear of the powers of Nature (or competing hordes) led to Humankind working on the highest level of social and technical improvement of individual safety over the last few thousands of years. From this aspect, western society lives at such a level of safety that has never been experienced before in history. We spend the bigger part of our lives in air-conditioned shopping malls, apartments and vehicles made almost totally independent from the influence of Nature. We only ever come into contact with Nature if we feel like it (e.g., we go on an excursion because the weather is nice). Western society has become such that most people live in peace—the safety of their bodies and material possessions is a given. So human beings enjoying western society's standard of living have reached the dream sought since the appearance of rational thinking: living in safety from other people and the dangers of Nature.

It is natural that people outside western society want to live according to the example of the 'ideal life' provided by the West. Nobody can be judged because their lives are led by the desire for the western standard of living—not even if it takes us to the destruction of Life.

But due to this millennia-old striving and the resulting technical development, we have literally become the rulers of the world. Our society's ability to change Nature is so powerful that we can almost totally rule Nature. It is also a fact that this 'ruling' can only take place for a limited time, but this is beside the point (although it is of particular importance as it can cause our destruction). Here is the connection between society and human frustrations: our society is based on the striving for an infinite amount of safety. The instinct for the pursuit of safety is an ancient instinct that is derived from our animal state and feeds on fear. As a matter of fact, we can say that our main way of thinking is led by animal instincts. We still live in a society that is driven by our fear of Nature and other competitive people. This has resulted in the natural consequence that society is built up in a way to protect people above all from Nature and others. We want to rule Nature and take control of other people to protect ourselves from the fears established in ancient times.

What was our initial assumption? It was that all the environmental damage and unrest is the manifestation of Humankind's spiritual problems. Now it can be clearly seen that this assumption is absolutely right. Our instinctive fear of Nature and our fellow humans is the basis for the system of our current society. In this way, society is inherently environment-destructive, as it does not want to live together with Nature but to rule it. Our society does not want to live together with our fellow people but to live in safety separated from them, or in pathological cases to control them. On top of it all, there is the environmental impact caused by the addictions present in the bigger part of society. As a result of this combination, we are depleting our natural resources in an unsustainable way, taking away the well-being of future generations for good—and we could even ruin our own near future.

The solution is multifaceted. Technological and scientific improvements are obviously crucial in order to live in a more environmentally friendly way. It is also important to change people's mindsets regarding their lifestyles. But this chapter is about to point out that until we start changing the root causes of the problem, we can only slow down the process of Life destruction but would be unable to stop it. We have to solve the problems which are causing our current destructive way of life. What are these problems? Humankind's spiritual problems and the social system based on them. I hope it is obvious and also proven to you that the most efficient protection of the environment is through our spiritual development, and the healing of our fears and frustrations. What is more intriguing is that this healing process is quantifiable both on the social and on the individual levels. I would like to talk about this in the following chapters.

2.9. Your thinking covers your body's wisdom

I had been contemplating for decades how it can happen that a dog who was separated from its mother at an early age and brought up alone knows exactly which type grass to chew if its stomach is upset. Or how it knows which puddle not to drink from even if there is an odorless pollutant in it. How does the carp in the lake know that the octopus-flavored bait is good for it when it has never tasted octopus before, or why does it like corn if corn doesn't grow under the water? The answers were given to me through the science of kinesiology. It was incredibly interesting when I started to read about it. I truly recommend the psychiatrist Dr. David R. Hawkins' book *Power vs Force*. This book provided the scientific basis for the chapters about the levels of consciousness.

The science of kinesiology started from the observation that body does not make mistakes and is not able to lie. This may seem strange or unbelievable at first but it is so. Several thousands of examinations were carried out in the 1970s and 80s, in the dawn of kinesiology, and they could prove their assumptions with 99.9% certainty. Kinesiological measurement is done using the body response method. In this method, the body's response reaction can mean only yes or no. So the disadvantage of the method is that we can only get the answers from our bodies' wisdom to questions to which it can 'say' yes or no. (D.R. Hawkins 2004)

In an experiment, there were 5 closed envelopes with a pill in each. Four of them were poisonous; one was not: it was vitamin C. To provide total correctness of the experiment, all of the pills were with the same size, weight, form and color in spite of the fact that the participants could not open the closed envelopes. Then the participants were asked to choose the envelope with the right pill in it. Unsurprisingly, they hit the statistics norm: about 20% of the participants chose the right pill and 80% of them chose a poisonous one. With another focus group the kinesiologists examined their bodies' reactions. According to their bodies' responses, more than 99% chose the right pill, which is impossible with rational thinking, but was true! Just as dogs do not make mistakes in which one of many types of grass they can eat, the human body does not make mistakes either in such cases. Kinesiology is a proven and accepted science today. I also can use in practice the body response method on which the whole science is based.

The dog and the fish differ from people in the fact that their rational thinking does not obscure all their gut feelings and impulses because they do not think. The dog and the fish feel what is good for them and what is not, and they rarely make mistakes. People lost communication with their bodies because of their rational thinking; they are not able to pay enough attention to their signals. In many situations though rationality can fail, and in these cases an animal is much wiser than a human. It's weird to think about it, isn't it? To our vanity it may be unacceptable at first, as we were raised to believe that we stood out from wildlife due to the ability of thinking. While this is true, we lost a wonderful skill in exchange for knowledge. I am convinced that if people's spiritual development increased, people in the future would be able to switch off their rationality when it was needed and as a result their ability to respond to their natural perceptions could be used again. But at present we are very far from this; our rationality covers up everything.

It is important to highlight that rationality is a very good thing. The problem with it in our current world is that we want to use it for everything. If, for example, I want to create new software that brings a better quality of life to thousands of people, I will program it on the basis of my rationality because I want it to work well. At the same time, how many situations you have had in your life when it was hard to force yourself to think rationally, and doing so caused big tensions within you? Or when in spite of rational knowledge you were totally helpless regarding your spiritual problems? In these cases, your body (and soul) knew the right answer—the inner tension came from suppressing your body's and soul's signals. My life has become much easier since I started paying attention to these signals, not just because the inner tensions have reduced in me and I live a more peaceful and happier life, but because in several cases my decisions which initially seemed like irrational silliness came out to be right. Some days ago, for example, I had a gut feeling to leave 15 minutes earlier on my way to a distant city than what the travel planner software suggested. I did not have a solid reason, but I listened to my feelings and left earlier. After about an hour there was an accident on the highway, causing a bad traffic jam. In spite of this, I arrived sharply on time for the meeting. If I had decided rationally, I would have been late for the meeting. This situation could not have been solved with rational thinking because when I listened to my gut feeling the accident had not happened yet. The world of intuition is a scientifically proven thing, some aspects of which were proved by the California-based HeartMath Institution.

There is no problem with rationality, as it is rationality that has given us technical and scientific development. The problem is with losing the right proportions. In western society we believe in the Holy Grail of rationality; however, it causes partial blindness. It will not be so **in future society: we will handle everything in the right place, both in the world of intuition and with rationality. It will lead us into a more peaceful and purer world.**

Here's some good news: while it's true that you cannot reach the deep, unconscious messages of your body by using your brain, a good kinesiologist can. The expert can see your body's responses to any questions with a simple body response method. What this method is good for? For plenty of things, **your body is much, much wiser than you are** (sorry!) and your body never tells lies (sorry again!). But it is certain that you lie to yourself, if not to others.

We lie to ourselves often because it is easier to believe other people's truth than to face ourselves. In other situations, we lie to ourselves due to the desire to fit in. It is really rare that someone can be honest with themselves. On the other hand, our body knows exactly what is being suppressed in it.

So, what is a kinesiologist good for? They can bring deeper self-awareness to open and release suppressed spiritual wounds. The extra knowledge that a kinesiologist has compared to a psychologist is that they always work from truth. Namely, during a traditional 'conversational' psychotherapy the patients describe things from a personal perspective that is more beneficial to them. There can be such factors behind their problems which they do not dare to admit to themselves—not even to their therapist. In contrast, **you cannot lie to a kinesiologist since they always work from the true body response.** This is highly important from the perspective of the reliability and measurability of the method.

Why is this useful regarding climate change? Because the science of kinesiology can quantify every person's actual level of happiness, and this in turn opens a door to us for solving climate change.

2.10. Quantifiable happiness - or the levels of consciousness

From the previous chapters, it should now be obvious that **while Humankind is in its current state of mind, climate change cannot be stopped;** only its progress can be delayed or its impact reduced. Even if you are not interested in spiritual matters I urge you to hang in, because only a small portion of the solutions will concern spiritual issues. Every item within the system aiming to solve climate change is absolutely rational—even the ones dealing with spiritual issues. My engineering-scientific nature simply does not let me search for solutions through means that have not been proven scientifically and rationally.

This chapter introduces a group of crucial basic notions on a basic level. Those who understand and acquire them will significantly change their view about people and how human society works. The broader vision to be gained is important for reaching new milestones on your way to happiness, and at the same time provides an important basis for understanding some tools in the fight against climate change.

The foundations of the system I am now going to introduce were summarized by Dr. David R. Hawkins in his book *Power vs Force*. It's not an exaggeration to say that this book belongs among the best pieces of literature dealing with the human spirit, and I wholeheartedly recommend it to every kind reader. If I was in a position to determine school curricula, I would include this book in the teaching materials for both primary and secondary schools, as it can fundamentally alter people's perspectives.

It has been scientifically proven (D. R. Hawkins 2004) that the human spirit has a particular energetic level. This energy has an impact on the people around us and reflects our inner mental health. Levels of consciousness can be associated with the individually defined levels shown in Figure 3.

The average person's spirit has already experienced most of the 17 feeling systems presented here. However, the average value of our every feeling provides our actual level of consciousness, which is characterized by one of the main feelings listed below. The level of your consciousness depends on which of these 17 levels your spirit stays in for most of an average day. Obviously, our mental state can be unbalanced by cases in which we are more enthusiastic and happier, or in other cases when we sink into a darker frame of mind. From the level of shame up to the level of pride, people radiate enough negative spiritual energy to make their lives destructive; they have a negative effect on the people and every living creature around them. At the levels from courage to enlightenment, the human spirit radiates positive and life-supporting energies with a proven, favorable overall impact on its surroundings. The level of consciousness you are operating at can be measured within three minutes with the help of the body response method as performed by any good kinesiologist.

Surely you have already experienced that being next to a spiritually mature person you felt more peaceful, happier and more energetic. And surely you have also experienced that being beside a complaining person

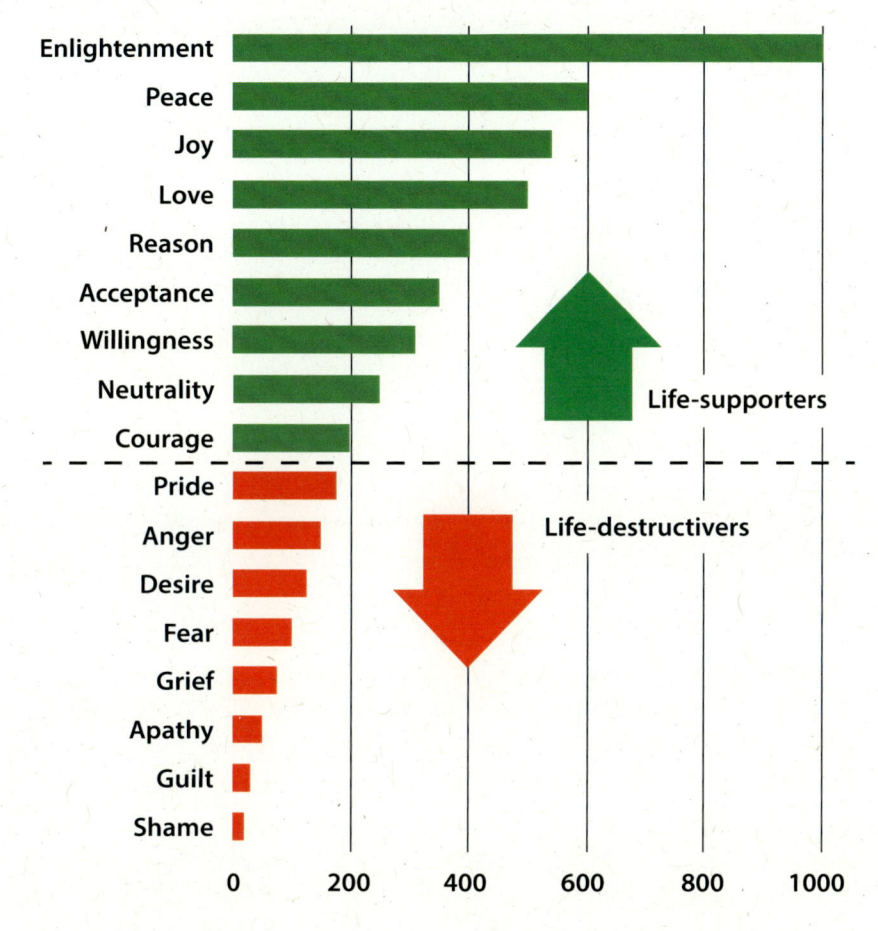

Figure 3: Levels of consciousness (David R. Hawkins, 2004)

who feels only self-pity, your energies decrease rapidly. These latter people are called energy vampires. Because the energies of the spirits have an impact on each other and they always strive for balance, two or more people close to each other have an impact on the energy levels of each other's spirits, even without communication. For instance, even if you are standing on the subway and caring about nobody, it happens to you as well.

If a person at a very high spiritual level gives a lecture, the audience in the room is usually taken by the positive atmosphere that is radiated by the lecturer without the audience being aware of it. Such a person can be on such a high level of consciousness that they are be able to compensate for more than 10,000 people's negative spiritual energy. Think about Buddha, Jesus or Mother Teresa: what an incredible impact they had on people in their surroundings! According to my interpretation, Buddha and Jesus were enlightened people, but I do not mean to challenge anybody's religion; the religious explanation and mine do not cancel each other—they rather complete each other and one makes the other more understandable.

If we look into our lives from the perspective of this system, we will get the answer to why it is so important to deal with our spiritual development. We affect other people's happiness beyond our own, without the given person being aware of it. The most important step in spiritual development is when we move on from the level of pride to the level of courage, since here we become a Life supporter from having been a Life-destructive Person.

When the average of Humankind's level of consciousness reaches the Life supporter level, environmental destruction will decline, the fight against climate change will be effective, and unrest will decrease. This happens because people will reduce their Life-destructive way of life instinctively from inner motivation. This is why your personal happiness and the question of climate change are developing from the same root.

But getting back to your life for a moment, if you want a happier life, a more peaceful family, or more devoted friends in your surroundings, you will have to raise your level of consciousness by developing your spirit. It is the greatest thing you can do for yourself, your loved ones and the world. This is why the wise ones say that happiness comes from within and is not the result of external forces. If you change the level of your consciousness in a positive direction, you will be more peaceful, harmonious and affectionate, and your spirit will not attract so many bad things. The quality of your personal contacts will improve, and what's even more wonderful is that there will be even smaller negative vibrations in your life. It's worth it, isn't it? I already can tell you from my experience that the answer is 'YES'. Some years ago I was unconsciously living with Life-destructive levels of consciousness. Now I live on a very high level of affirming Life, as proven by a kinesiologist's measurements. I do not mean to brag about it, but I'm sharing this success because **if I managed to achieve it, you will manage it as well** if you are open to it. The consequences are almost incredible. There have been quality changes in every area of my life. If I had to characterize my life five years ago with a single word, I would use 'suffering'. If I had to do it with my life today, the word 'wonder' would be the most suitable one.

It is important to know that you perceive the world differently from every level, and your vision about truth is always different as well. If you understand this, you can be much more empathic with other people as you will understand how others think so differently about things than you do.

To recall and re-experience these emotions in my life brought me closer to understanding people. How many people must there be in the world who are suppressing similar feelings of suffering within themselves forever, carrying them deep inside their souls and playing their ever-crazier games due to it unconsciously?! Traumas from childhood, divorce or losing parents, losing attention due to another sibling, parents dragging their souls through the mud, deprivation, being exploited… these and similar deep spiritual wounds burden billions of people. This realization helped me recognize that we **cannot judge other people because we will never know why the other person is the way they are.** What is a more interesting and exciting thought is: if I had been in the other person's life, supposedly I would behave as they do now. Jesus Christ referred to this when he said we should forgive those who trespass against us, as they do it unconsciously because of the suppression of their damaged spirit. If I fight back, I will just deepen their wounds. But if I forgive them, I will help them with the healing of their spirits. Obviously, fulfilling these words is impossible under a certain level of consciousness. There is such a huge spiritual burden weighing on us that we rarely have energy to deal with others, and we may react to the smallest attack with huge, sudden confrontation. The world of regular forgiving starts above the Life-supportive level of consciousness of 200.

For further exploration, I've written about the different approaches for connecting to the levels of consciousness in Appendix 3. In the following section I am going to briefly discuss how differently people see the question of climate change depending on their level of consciousness.

Shame and guilt (energetic level 20 or 30): The level of Shame and Guilt is the deepest, blackest, darkest feeling that a person can experience. Here you are so close to the level of death that you see yourself as being zero. In most cases, people at this level truly desire death. Unfortunately, the feeling of being worthless often provokes scornful and hateful reactions towards the world. Watching the world through this miserable lens makes Life seem vicious, and we imagine a really dark future. People living on this level think that climate change will definitely destroy the world, and that Humankind deserves it.

Apathy (energetic level 50): The person living at the level of Apathy is the one who is not interested in anything anymore and nothing matters at all. They feel this world can give them nothing that can be valuable. There is complete hopelessness at this level; our future vision is really pessimistic in this state and we also see the world's future as being dark. To a person like this, climate change is equal with dying out.

Grief (energetic level 75): At the level of Grief we are despondent spiritually and we are prone to being stuck in the depths of self-pity. We see our life as tragic and reject help from our surroundings: 'They don't know what I'm going through.' In this state we do not have the energy to do anything for climate change, as we are not even able to do enough for our own lives.

Fear (energetic level 100): It was mentioned before that our whole social system and the bigger part of our thinking processes are based on fear. Now that you can see what a Life-destructive level it is; it should be clear why the solution of climate change is impossible without healing our souls and changing our thinking processes. Our soul is at this level if most of our everyday thoughts are penetrated with fear. The climate-hysterical people belong to this level. They are terrified of the future caused by climate change or of a nuclear disaster.

Desire (energetic level 125): as this level is not so low, hope becomes strong and our spirit is beginning to long for good outcomes. Desire gives us the strength to set out to reach our goals. At the same time, this level of consciousness is a rampant world of addictions. People living at this level suffer from one or more addictions; it is consequently a heavily Life-destructive level of consciousness. The fight against climate change is an incredibly huge fight from the perspective of people living at this level. If there is solution at all, it is very far, very remote, and can be only reached through a vast concentration of forces and collaboration.

Anger (energetic level 150): At this level of consciousness the main emotion saturating the soul is hatred. People living at this level always search for a target (a target person). They often find themselves involved in fights or are part of aggressive scenes. They are the permanent troublemakers of clubs. Most of the climate change activists who try to impress the world with violence, arguing or harsh words in order to change are on this level. It is also characteristic of most on this level to blame multinational companies, airlines or others on climate change and to express it in a rude way. They are the driving force of the unrest in the world without being aware of it. As they constitute a big number in our population, the unrest is big as well.

Pride (energetic level 175): this level of consciousness is typically the world of selfishness and tough ego. Because the world today always suggests that we should be selfish and we should deal only with ourselves, most people live at this level. Due to the fact that it is a Life-destructive level, never-before experienced environmental destruction and ominous climate change prognosis are the indirect consequences of it. With the strengthening of selfishness, real communities are falling apart. Everybody cares about themselves and expect every other person to care about them as well. There is hardly a chance to form real and deep relationships. Of course, it is always the other person who's to blame if the relationship didn't work out. At this level of consciousness the individual does not realize that the more selfish they are, the further they get from real and quality relationships, and the more they destroy the Earth's natural resources, apparently. A selfish person wolfs down and experiences everything they desire! They do not have enough self-control and they do not even need it. For people at this level, the point is to live every experience, reach every desired goal and possess material things. A person living at the level of Pride does everything only because of self-interest. Their only motivation is themselves and reaching their own targets. The root of their Life-destructive energies is exactly that they are not able to place the social, ecological or other public interests before their own, and because of their selfishness they forget about the real target of their lives. They can identify with these targets only if they correspond to their own personal targets. They always believe in what strengthens their self-interest and self-image, and they want to persuade the people around them about it. That is why these people like to point out arguments which prove there is no climate change or that it is not a big issue. This way, they do not have to confront their conscience. These people deeply feel that they deserve what they desire. They do everything to reach their targets; they are even ready to bend the rules in many cases. If they cross a line, they justify the exception and claim it was the right thing to do. As with many other Life-destructive levels, it is also fed by the deeply suppressed roots of fear. Due to this fact no one who lives at this level can be an efficient climate protector. However, once climate awareness is built into their ego, they can be a loud advocate of climate protection and because these people are often powerful they can impress others in this field.

Courage (energetic level 200): the level of Courage is where the spirit rises to become Life- supportive. Significant quality changes happen in a person's life when they step from the level of Pride to the level of Courage. And therein lies the solution to climate change: as many people as possible have to cross to this level! The good news is that it costs nothing; economic power is not needed! Only You and the intention to be happier are needed! Compared to the Life-destructive levels discussed so far, this level is where the individual first truly faces themselves. This is the most important change compared to the previous levels: these people can hold up a real mirror to themselves and start to see their own mistakes, weaknesses and frailties realistically. They can then start to face how climate-destructive their lives are and how much must be done to change it. An incredible change is that the individual finally takes responsibility for their deeds and doesn't make excuses or blame others. It becomes clear at that point that it is not the multinational companies who are responsible for climate change, as 7.9 billion customer decisions keep them alive. If I change my own life, I will weaken the system. And if I tell others about my decisions and actions, I will weaken the system more.

Neutrality (energetic level 250): At this level there is no more hurting others or exaggerated self-assertion. When I arrived at this level my view about the world and future and my approach to other people changed immediately. I also started to feel some really deep and calming peace and harmony inside. This kind of spiritual sense of safety had been totally unknown to me before. My vision also altered from pessimistic to optimistic. If only everybody lived on this level! If it were so, there would be world peace, and environment pollution would decrease to an extremely low level. Why? The answer is easy: both depend on the mental health of 7.9 billion people and many other things as well. I hope this fact is beginning to take shape within you too. At this level, the main developmental process of our spirit is release. Here we put down the dragons of our minds and the addictions which had been pulling us down. It suddenly releases a huge amount of Life-supportive and selfless energies in ourselves. Writing this book is also Life-supportive and selfless; however, it is a really good feeling because that huge amount of positive energy wants to break out and do good to have a better world. A leader at the level of Neutrality believes in the success of mutual dependence, and due to this they attain greater successes. (Mutual dependence is a crucial basis of the solution to climate change, so I wrote a separate chapter on the topic.) The basis for managing a high level of teamwork is that I trust in my teammates' decisions. This success cannot be measured financially but rather in the social usefulness of the team guided by it, or in the spiritual and professional development of the members of the team. At this level the individual starts to trust in people's goodness and in a positive future. This is a new perspective, as on the levels discussed previously there is neither the strong trust in the future nor the belief in the goodness of people. The belief that there will be a way out of climate change appears at this level. We start to believe in the power of human cooperation and creativity. It is a wonderful level of consciousness. Unfortunately, most people living today never reach this level. Why? The answer is pretty simple. Instead of spiritual development, most people spend their lives with formalities. Chasing money, a career and an attractive look are not bad things, but they do not lead to happiness at all. This condition is the level of people who have finally stepped onto the island of happiness. As I entered this level a few years ago, I know how wonderful it is and I would love for you to experience it in case if you are not already there permanently. This way of being can do the most for your happiness and beyond to the Earth's future.

Willingness (energetic level 310): a person living at the level of Willingness differs from the previous levels mainly in being basically optimistic. Such a person cannot be easily dislodged from their optimistic perspective. It becomes instinctive here to see the glass as being half-full, and if anything bad happens we usually draw positive conclusions. There is a particular intention in the individual at this level to develop further in the spiritual world. At this level we want to give up all of our inner spiritual processes that cause negative levels of conscience deliberately. It becomes clear to us here that happiness comes from inside and is almost totally independent from every external impact. At this level our vision is hopeful and optimistic. It is always the search for solutions and a trust in Life that characterizes the individual's view regarding the future. People at this level who take the problem of climate protection seriously try to actively act against climate change with all of their power. In order to advance our future in the right direction I adjusted my lifestyle, shaping my family life, giving lectures at the university, building environmental awareness into my companies' projects, giving examples, writing my blog and writing this book. I also have several other ideas. I hope I have the power to carry out them.

Acceptance (energetic level 350): The consciousness level of Acceptance is a wonderful and happy world. Accepting ourselves brings the acceptance of our environment. We will become forgiving with the people in our environment and we will accept the world as it is. Of course, this does not mean we do not want to improve our environment. Instead, it means that we can use all the energy freed up from pointless opposition to really change the world in a positive direction. I am frequently asked: 'How do you have energy for this book in addition to so many activities?' The answer is simple: all the energy I used to use to fight against myself and the world has been released. At this level of consciousness, harmony is our outlook on life that penetrates our way of living, such that the inner fights decease and forgiveness is the main emotion in addition to acceptance. We forgive ourselves every sin and bad deed of our life so far and we do everything to make up for them in the future. We realize at this level how many wrong things we have done so far in our lives. At the level of pride (or any of the other Life-destructive levels) we simply do we recognize how many bad things we do. We destroy our environment unconsciously while being convinced about our personal truth. At these levels our ego covers up the consequences of our deeds from us and idealizes why what we did was right. At the level of Acceptance we truly feel and see how many wrong things we have done and still do. Here we take all the consequences of our deeds and do everything we can to make up for them. If 20% of Humankind reached this level, there would be world peace and environment pollution would be reduced to such a low level that the destruction of Nature would stop and Nature would start to recover globally. As I have been down this road, I know it is possible. You too can step onto this road and reach the finish line.

Reason (energetic level 400): The main emotion at this level is understanding, as at this level we are able to think beyond our own problems and perspectives and are able to understand others entirely. As a result, people living at this level are able to identify the world's processes that are incomprehensible by others. It is not by accident that most Nobel prize researchers, inventors and famous thinkers or philosophers live at this level. The main point of these people is their outlook on life. They feel they have received their abilities to show the right direction to society with their example and significant results. The individual living at this level think of climate change and unrest as problems to sort out, and they trust in human creativity and knowledge to find a solution sooner or later.

Love (energetic level 500): The person at this level understands that selfless love can come through everything; it is the main power of the Universe. This 'understanding' in an individual means that they feel the existence of love in everything everywhere with complete empathy. The one reaching this level is able to feel the 'radiation of love' of the whole Universe. They do not need to get feedback regarding love from other individuals to able to feel they are lovable. Mother Teresa lived at this level (D. R. Hawkins 2004). These people spread so much love from themselves that it is contagious, and simply being in their proximity generates wonderful feelings in us. The whole person is permeated by a harmony that you feel instinctively in their presence. The sense of unity with every living creature appears in them more frequently. As a result, this kind of person is not able to hurt a fly. They love and respect every living being infinitely. However, the average person is very far from this, but if they are at this level in the far future, Humankind will live in world peace, in perfect harmony with Nature. There will be no selfishness, aggression, or destruction of environment. This book was also written to convey that spiritual development is the most important pursuit not only for ourselves but for the fate of the world. Followers soon gather around such people, who spread their news far if they are open to it. However, as modesty is a basic characteristic feature at this level, I am convinced that there are a lot of people living at this level of consciousness living in seclusion, without being aware of the power they spread which dramatically slows the destruction of Earth. Of course, most of these people do it selflessly and unconsciously since they do not possess the scientific inventions of kinesiology mentioned here.

Joy and peace (energetic levels 540 and 600): The person living at the level of Joy experiences the sense of their own wholeness. This means that they have accepted themselves so perfectly that they find nothing in themselves that cannot be loved. At the same time the inner attitude radiates from within as well. Such a person experiences the wholeness of Life. This level of peace means the individual experiences the wonder of Life's perfectness inside and outside. They reach this level through both self-acceptance and accepting the world. People living at this level are able to neutralize the negative energy of hundreds of thousands of people who live at Life-destructive levels. If such a person lives in a city, people living in the same city are more peaceful without being aware of the fact that it is because there is such a person nearby. There are fewer accidents and less criminal activity in such cities. People living at this level are steeped in a sense of joy. People living at the level of Peace live in the feeling of gratitude. They experience themselves and the world as being blessed, and this huge peace experienced in spiritual safety is the strong base of their spirits. That is why there is no nervousness, stress, fear or other forms of spiritual pain. Regarding their outlook on life, these people are characterized by infinite goodwill and living the wholeness of Life. They place all their deeds at the service of Life and in this way they do a lot to save Humankind and terrestrial Life. However, they are extremely humble so in most cases these situations are not known.

Enlightenment (energetic levels 700-1000): we have come to the last part of the series of writings about the levels of consciousness and to the highest level of human existence at the same time. At this level everything becomes perfect. Our consciousness perceives Life and the world as a shiny wonder. The spirit at this level is perfectly clean; consciousness does not connect any thoughts, feelings or features to anything. Those whose spirits are at this level live in the sense of pure existence. You just simply are, as part of a perfect system called Life, and where you live is in unity with every other thing. It is such an uplifting and perfect feeling that it is hard to find a word for it. The harmony made up of peace, joy and selfless love almost overflows. It is incredibly rare: one out of hundreds of millions or of a billion people stays permanently at this level. At this level of consciousness, existence itself becomes the meaning of Life. Existence itself is the aim.

The people who lived at this level in the past had an enormous effect on the world. Such a person's level of consciousness is able to compensate and neutralize several million people's Life-destructive levels. Miracles become commonplace at this level as the spirit is on a such energetic level that the individual is able to do things which seem to be impossible with just logical thinking. The founders of big world religions and world-changing wise people lived at this level, like Jesus, Buddha, Krishna, Lao Ce, etc. They are the people whose ideas will last until the end of the world and will have an impact on Humankind (D. R. Hawkins 2004). If about ten thousand such people lived on Earth now, they would solve the problem of climate change and unrest in place of the whole of Humankind.

Now that you see the steps of the development of levels of consciousness, it should be clear to you which is the direction our evolution leads. I now know that the aim of human evolution is to develop our spiritual consciousness. This is the only way that leads Humankind to the world of global peace and harmony with Nature. Otherwise, we will die out; it is just a matter of time. As we live more and more consciously from generation to generation, the average person gradually rises to higher and higher levels of consciousness, and enlightened people will turn up more and more often in the world. Meanwhile, Humankind's average level of consciousness will rise and environmental pollution will decrease, world peace will grow, and individuals' average happiness level will rise. If you think about it, this has been happening from ancient times until now. The next chapter is about this.[4]

---

[4] If you want to read more about the levels of consciousness, I recommend appendix 3.

2.11. Group dynamics and some conclusions

The level of consciousness is a special type of energy. This kind of energy can be felt outside the body. Moving away from the body its power is reduced, but it depends on the individual how far their presence affects others. The system built up by this kind of energy works like every other energy system in the Universe: it strives to achieve equilibrium. Think it over: the fire goes out, the wind quiets down, all the water flows to the sea. All of these are natural manifestations of the pursuit of equilibrium. In order to understand how the energy system of the level of consciousness works, here's a simple example. You are just two people together in a square—no one else. Your level of consciousness is 270 while your lone friend's is only 30. You are talking in the square. What is the result? The two people's level of consciousness affect each other and the created energy system strives towards equilibrium. At the beginning of the conversation your level of consciousness starts to decrease while the other one's starts to increase. If the function of the relationship were linear, your level would decrease to the level $(270+30)/2 = 150$ while your friend's level would increase to that value. (The function of relationship is actually logarithmic, but I don't want to scare you with such harsh topics). So let's just say that equilibrium is reached at about 150 (in reality it is much higher). This does not happen deliberately; it is an automatic process that works naturally, and neither of the two people can affect it. They simply feel the following: the one living on energetic level 30, who suffers from guilt and has a difficult life, feels themselves becoming more energetic and getting involved in the conversation more enthusiastically as their spirit's energetic level begins to rise due to your impact. Your energetic level will begin to reduce, though. You will be restless at first, then increasingly distant when your energetic level goes below 200 as you are now at the level of pride. Then when you reach the level of anger at the same time (150, balance), a quarrel breaks out and both of you leave the square. Your friend falls back to the level of guilt and feels an even bigger guilt over why they caused a quarrel. Your friend will blame themselves and will hate themselves because of it. They will sink even deeper in their current condition and draw the conclusion that the world is evil; their misconception will affirm itself. You, out of the quarrel, are facing a confusing situation: how could you get into such a big row over such a silly thing when you are basically a really peaceful person?! The thing you really do not understand is why you feel squeezed. Your peace of mind will return slowly and you promise yourself the next time you see this person you will try to apologize for your behavior and clarify the problem, as you are on the level of neutrality—a pretty realistic mindset.

You can now clearly understand how energy vampires work, can't you? They do not drain your energy consciously: it is the result of their low level of consciousness. They feel better next to a person on a higher level of consciousness, and as a result they like to meet people like that. In contrast, those whose energetic levels are reduced want to avoid people who presence causes this.

Imagine that every person's spiritual energy affects others around them in this way, regardless of whether they speak to each other. If for instance two people start to argue on a noisy bus but you can't hear what they are talking about and are not even aware that they are arguing, the tension will start to increase within you regardless. Your level of consciousness will have started to reduce in the direction of the level of anger.

There are of course more positive examples as well. It has also happened to me several times that I felt more enthusiastic and balanced while talking to somebody. There were cases when every student was fascinated by the heat of my speech and they hardly realized the time passing by. Once I had the opportunity to mediate with a Zen master and experienced the deepest meditation experience of my life, and not by chance, as his high level of consciousness raised me to a height where I have rarely been in my life so far. **The power of group meditation and communal life lies in this. Communal life is a crucial tool for stopping climate change!**

So regardless of whether you speak to other people around you or not, they affect your level of consciousness and you affect theirs as well. The exciting part is that people on extremely high or extremely low levels of consciousness can compensate for many other people's levels.

This is why Jesus, Buddha and Mother Teresa could affect so many people positively, and why Hitler and Stalin could affect so many people negatively. People on extremely high or low levels of consciousness are able to reduce or increase several hundreds of thousands or millions of people's level of consciousness because of the logarithmic function relations. If you, for example, reach the energetic level of 350, you can neutralize the negative energy of about 10,000 people who are at the level of Pride. This is why the example of you and your friend in a square was an oversimplified one. The balance will not be 150 in reality, and this is the most intriguing part of the whole question. **The most effective thing you can do for your happiness and for climate protection is to raise your level of consciousness. What is more interesting is that this is the way you can do the most for your fellow humans too.** If your level of consciousness is high and you live in a block of flats, the other residents will not know why they feel more peaceful when you arrive home. They will think it is simply the warmth of home while in fact it is the effect of your spiritual energy from two floors below. In this way, if your level of consciousness is high, you will raise the average level of consciousness wherever you go while helping other people, unbeknownst to them. It's the same with your family. You come home and bring peace and harmony with you. It is incredible how wonderfully effective this whole thing will be on the lives of you and your loved ones. And you do it selflessly; others do not have to know what's going on—it simply happens day by day. This is the real protection of Humankind and climate, so please think of it when I say: **be happier, raise your level of consciousness and save the world with it!**

    I suspect a question has already arisen in you by now: how can I raise my level of consciousness? Unfortunately, there is no place to fully address that in this book, but I am already writing a book which can lead you on that way. I hope you will honor me with your attention when it is published, but I can give you more ideas in the meantime: read the book Power vs. Force by David R. Hawkins. At the same time, in Appendix 3 you can find more details about the levels of consciousness.

2.12. The ignored evolution

The levels of consciousness are not applicable only on the level of individuals, as we have already discussed. If we look at Humankind's current average level of consciousness, we get the level on which Humankind actually lives. Currently this level is about 180-190, as determined by David R. Hawkins. This means that Humankind stands at the biggest threshold of great changes in history! But before going forward, let's look at this process from the perspective of human history.

In ancient times, humans were afraid of the powers of Nature as their thinking encouraged them towards understanding but their knowledge was insufficient to reach it. This generated inner fear due to the lack of understanding of Nature, which is one of the deepest roots of our soul. A beautiful Bible story becomes understandable, namely when Adam and Eve took away the apple of knowledge and were taken out of paradise. While we were living as animals and had only our instincts we were not afraid of Nature itself—only some direct effects of it—but adapted instinctively to it, and were part of it. The experience of unity is a basic sense in the animal world. We sank into the world of constant fear by getting the opportunity of thinking from evolution, which was a great gift but we have not yet learned how to use it properly. In those times, the average consciousness level of Humankind would have been around 100.

Developed civilizations started to rise from the level of Fear, and the peaceful periods of some city-states in ancient Greek democracy rose extremely high to the level of Courage, whose energetic level is 200. At the same time, the famous Spartans reached the level of Anger (150) during their development. The average of the human world at that time would have been much lower. Unfortunately, there was a dramatic setback in Europe's dark medieval times. The Christian church and the operating feudal power in cooperation held the tool of fear over Humankind. The 'developed' part of the world was ruled by the fear of divine wrath and going to Hell, which resulted in Humankind falling into the Dark Ages. It is not by accident that this was the time when we destroyed the South American cultures which had been on a really high level of consciousness compared to us. The Incas from these cultures would likely have lived at the level of Courage. Their peaceful, Life-supportive world was literally destroyed by the aggressive, European 'culture' living at the level of Fear. It was a devastating loss for the world[5]. The North American Indians and people living in Japanese samurai culture also lived at a higher level of consciousness than we Europeans lived: they lived at the level of Anger. This is still Life-destructive, but it is still a much higher level than the level of Fear. The world average was likely reduced to the level of 110 due to the horrible acts of white people.

The 'developed' part of the world stepped up to the level of Desire (125) due to the Industrial Revolution and the Enlightenment, which was a driving force in scientific and technological development. At the same time, the eastern cultures (Buddhism and others) in the colonies with much higher levels of consciousness started to sink to our level due to the oppression and the torsion of western 'civilization'. The Life-destructive forces of the world were slowed by the Native American and Eastern cultures, however, with the total destruction of the Native American cultures and the pollution of Eastern cultures we pulled down the average level of the world's level of consciousness to our lower European level. It's no coincidence that white people are instinctively not liked in several parts of the world. Unfortunately, from the spiritual point of view we have done only bad things to the world. It's time to make it better! **That's why in the fight against climate change, western economies have to take more responsibility.** Our narrow mindset and primitive spiritual development got the world to where it is. Obviously, a vast amount of courage is needed to face this fact, accept it, and start to make the consequences of our deeds better. It is no surprise that Courage is the first Life-supportive level of consciousness.

By the beginning of the 20th century we stepped from the level of Desire to the level of Anger (energetic level 150). The eastern cultures that had survived up to that point sank to this level while we Europeans rose to this level. Desire grew so big in some nations that the national interests started to battle. It was unavoidable that anger spread everywhere and

---

5   However, it was not completely a bed of roses; just think of the human sacrifices

finally exploded worldwide. Even if this seemed to be a simple huge conflagration, in fact it was an indispensable step from the perspective of the development of Humankind's level of consciousness. The energy of Anger disappeared in the two World Wars, and the western part of Humankind reached the level of Pride (175) by the 1960s. Meanwhile, western civilization completely polluted all the other cultures of the world. Capitalism and the dependence of intangible assets became a global cultural structure whose last big barrier—socialism—was consumed by development.

At the same time, the rise of the western world to the level of Pride made the other parts of the world sink to the levels of Desire and Anger. People living in poor countries are longing for the quality of life seen through global media. In some countries this craving overflows in such a way that wars and migrations are started, which we are experiencing today. It is also not by coincidence that Islamic extremists living at the level of Anger consider European civilization to be their main enemy. We brought this on ourselves; it is a consequence of our historical deeds and an indispensable stage of Humankind's spiritual development.

The world's average level of consciousness is about 180-190. Humankind has gone through incredible development, although a big part of Humankind still lives at a Life destructive level. **Level 200 is the level of Courage. By crossing this threshold Humankind would move into an unprecedented great happiness and world peace.** However, it is not sure that we can cross this threshold. Global climate change and the huge social tensions that are caused by Humankind's Life-destructive level of consciousness could destroy us before we are able to make this leap. The path we can take to move on to another level of development is a really narrow crossing. If we die out, after millions of years another species will appear that is able to reach that level of development and might be wiser than we are. If we do not die out, we will become that species. The stakes are high whether we find this path or not.

The problem is that western civilization's examples made Humankind blind, and the gleam of material things eclipsed its relationship with its own soul.

The Pope also pointed this out in his call regarding climate change (Pope Francis 2015).

This global blindness has to cease in order to make the rediscovery of our conscience and spiritual development the central part of our lives. The way is easy: we have to draw people's attention to it, from individuals to communities! The unprecedented important task of the representatives of big churches and the spiritual leaders of the world is to help people in this direction. I really trust that it will succeed, as **this is the key of our survival.**

I have already mentioned the new level of evolution. Now you can see this topic in a more complex way as you now have the basic knowledge about the levels of consciousness. If we make spiritual development the main aim of our life, environment pollution and hostility among people will decrease dramatically and the average level of human happiness will grow. If we develop in this direction, humans will be the first species in the history of Earth's evolution who is able to find balance with Nature at its own discretion and consequently protect itself from dying out.

Just think about it: **if Humankind's average level of consciousness rises by 10%, we will survive from the scary effects of climate change and reduce the chance to having World War III to almost zero!** In other words, you can do an incredible amount if you raise your own level of consciousness. Also consider that if you rise to the level of 350, **you might be able to compensate for the Life-destructive energy of 10,000 people who live at the level of Pride. You alone can do a huge thing** to make this world better! Goodness is just such an enormous power. Now that you know it: the excuse that you are too weak and small to do something for the world is so improper, yet how many people say this when it comes to fighting against climate change?

This process of development has been under way since the beginning of human existence (so it is not new). Darwin's theory of evolution is great but it is not complete. At a certain development level of a species the frontal cortex appears, as it was with humans. From that point the direction of evolution will be completed with spiritual development alongside traditional genetically-based development, resulting in the conscious or unconscious rise of the level of consciousness. As the development of the brain provided the possibility for its development, it can be explained with genetically-based evolution.

At the same time, Darwin's theory of evolution has to be completed with this addition, and the big picture will be complete for us. What is crucial to fix here is that Humankind can be the first species who is not only able to mitigate its own needs consciously in adjustment to its environment, but **can give up its fear-based evolutionary functioning, which is characteristic of an unconscious species.** I have introduced the conscious base of our fear-based social system which is crucial from the perspective of finding a solution.

The other extension of Darwin's theory of evolution has been mentioned by several ecologists, and this is that the driving force of evolution is not competition but cooperation. The aforementioned Bruce Lipton argues the same. This approach, which is gaining ground in the science of ecology, leads us back to the respect for the difference between Nature and People. **Cooperation can be the only long-term survival and development strategy on which the solution system of this book builds.**

Regarding climate change, it is crucial that Humankind's average level of consciousness is currently at about 180 to 190. This level is still Life-destructive, so Humankind is emitting Life-destructive energy. There's no surprise that we're experiencing unprecedented environment destruction around us! However, we are close to the threshold of the Life-supportive levels of consciousness (energetic level 200), the destruction is gigantic as a result of the incredible number of people on Earth, and the efficiency of our devices is formidable. As a result of these factors, environmental destruction has accelerated to such a level that has never been experienced in terrestrial life since the dinosaurs died out.

The good news, however, is that Humankind is just a hair from the lower limit of the Life supportive level, whose energetic level is 200. If Humankind moved up only 10 to 20 points, the further destruction of the environment would first slow down and then stop, and trends would be reversed around the world. If we reached this stage, global warming would slow down, the reduction of natural areas would be mitigated, and Life-supportive processes would be dominant in the world. Nothing else is needed; people should focus on their spiritual development more than ever. There is an incredibly wide range of methods for developing and healing our spirits. The best news is that money is not needed to do it: only the change of your mindset.

Level 200 is the level of Courage. For an individual to enter this level means that the individual faces their real self, faces the real consequences of their deeds and does not escape behind the 'defense line' of ego. This is the task of the whole of Humankind! We have to face ourselves, our wrong way of life, our wrong economic and social systems and their harmful consequences. The level of Courage further means that we do not simply face these things but we are brave enough to realize our real frailty, mistakes and sins, and to purposefully start to make them good.

The good news is that incredibly many people stand at the threshold of the level of Courage! Those who have been living at the level of Pride permanently feel it deeply that something is not okay. Their lives are not right. I am already over it, but I can clearly remember how I was suffering before crossing that wonderful limit. My ego tried to explain to me in vain how cool it is. I felt myself uncomfortable like a 'fish out of water'. I was fidgeting but it was no good. If the many people living at the level of Pride woke up, Humankind could cross that invisible but significant limit within some years and reach the level above 200. This does not mean though, that only those people who live at the level of Pride have to develop spiritually! As we are talking about Humankind's average level, those who live at lower or higher levels also have to develop in order to raise the world average. **The individual's moral duty is to take their own conscience and spiritual development into the most important center of their lives! This is the most efficient way to save the Earth and furthermore to raise their own happiness!**

2.13. Rationality versus heart-coherent objectivism - or the exaggerated rational approach

I have already suggested many times that to change the world we need new cognitive mechanisms or to rekindle forgotten patterns of thought. One of the biggest barriers to personal happiness nowadays is the excessively materialistic approach: only those things that we can prove rationally exist in the world. We make ourselves and fellow humans believe in them. Love, intuition, and gut feelings are difficult to explain in a rational way, but they exist, and science can already describe them quite well. Since we know that material is also energy and there are more dimensions than four, science and rationality are heading in the same direction. Indeed, there are more and more fields where science and spiritualism have already connected. We can read extremely exciting scientific discoveries almost every day about questions of intuition, emotions, meditation and the afterlife. We live in fast-changing world where the old over-rationalized walls are falling down. This change has already begun. **In future society, rationality and spirituality will be hand-in-hand cognitive mechanisms, and people will always be able to activate the proper one.**

People basically think that they believe in the existence of something only if they have experienced it or heard it from somebody who is credible. It is really important to get rid of our narrow mindedness and **believe that what exists is not only what we have experienced.** Just because I have not experienced something personally does not mean that it does not exist! I know that this approach is much more difficult because it stands in contradiction with the instinctive theory of our brain to strive for safety. It means stepping out of our comfort zone. But we should admit that for most people, happiness begins outside of their comfort zone! How easy is it to live, stubbornly insisting that the only things that exist are the ones we have already experienced?! This narrow-mindedness is fed from the fear deep within us. But just think about how small the amount of our personal experiences and knowledge is compared to everything to the Universe.

Getting back to the materialistic approach, its biggest disadvantage is that during our upbringing and education we turned away from spiritual questions. The person of today knows much more about the physical facts of the world than any earlier generations, but is more blind regarding emotional and spiritual knowledge than people have ever been. We are trying to find solutions to our spiritual problems by insisting on using our rational thinking in a stubborn way, which is why there are relationship conflicts, a lot of divorces, social tensions, and plenty of children lacking love. However, you also understand now that **this is the biggest environment pollution.** There is often inner tension between our spirit and rational thinking—not to mention that Humankind has never been so lonely spiritually than now. Our implicit trust in rational thinking simply makes us blind to see what we can easily solve with our gut feelings, emotions and spirits.

As an engineer, I was taught at university about the total authority of rational thinking, so I highly respect and practice rational thinking and a materialistic approach in my everyday life. My whole profession builds on these, and they steep all the branches of development. It is crucial to understand that I am not criticizing them! I am asking you to **strike a balance between rationality and spirituality.** Be open to the irrational world since it is much larger than the rational one. Now I can clearly see that rationality provides only narrow and partial skills. Furthermore, I see exactly that **happiness cannot be found by rational and materialistic approaches exclusively.**

When rationality and spirituality strike a balance within ourselves, we achieve a much more interesting way of life than being either extremely rational or extremely spiritual. This is the world of heart-coherent objectivism. I am going to share my related experiences and hopefully help you to turn towards new more effective patterns of thought.

For long decades I was living in a world of extreme rationality, and that was one of the reasons I tried to make my life better at low levels of consciousness. As a result of raising my level of consciousness, a lot of exciting doors opened in front of me. One such breakthrough experience was when I looked at a disabled person and I saw them beautiful. Before then, I had only been able to see the rational manifestation of the body and relentlessly judged everybody by their appearance. I was critical of people and of course of myself. Until the age of about 45, I lived my life almost completely this way. It's frightening how blind and narrow-minded I was, but what's more worrying is that I thought just the opposite about myself: I fully believed that I was objective.

Since then, I have been able to see the limit of being purely within the rational world every day. It is incredible how people make their decisions on a daily basis in a narrow-minded way with the belief and hubris that they are making the best decisions possible, while they run into their unhappiness blindly and willingly. Furthermore, I have witnessed in the last months how unable the rational world is to save the life of someone ill with cancer. How weak this world is—how incapable—if there is something which cannot be described mathematically in an exact way. However, cancer is nothing but all the shame, guilt, lack of love, and lack of self-acceptance piled up in our spirit, which had been radiating so much negative energy over the years that some organs in the body started destructive processes against themselves. The cells give up the belief that they can create a unity with the body, and they separate. The same thing happens within the body fighting with cancer as in society today. The parallel is not by coincidence. The separation and polarization broaden the process of falling apart. One day I was watching an anorexic girl who was walking in front of me. It was amazing to contemplate her spirit. She was so unable to accept herself that she shrank and consumed her body until she became emaciated. The lack of self-acceptance in her spirit consumed her slowly either way. Cancer is the same, but is even more insidious than anorexia, as the emotions suppressed over many long years are often so deep that the ill person is not aware of them anymore. They do not dare or are not able to look into their spirit under the thick armor which they created due to self-protection. But the reason is the same: a lack of love and self-acceptance—the total lack of the feeling of unity.

When I see a smiling, kind and really overweight person, I see a similar thing. Their spirits are so sensitive that their body literally grew a thick layer to protect themselves. Constant smiling is also a self-protection against spiritual pain. The body is capable not only of reducing itself due to a lack of love, but also of multiplying its weight. The roots and reasons for the body's expansion and shrinkage are the same. But how does this connect to the dry and harsh world of reality? Before we move on to this, please think about how many times it was hard to listen to your own mind—how many times you had to literally force your spirit to listen to your mind? It was really difficult, wasn't it?

I would like to show you an example of the Achilles heel of harsh rationality. In the 1970s in the dawn of environmental protection, the impact on Nature of the electrostatic fields formed around electric power cables was widely investigated. It became realized that such a constant electric field could be the cause of several cancer diseases and severe changes. This is one of the reasons there are now clearance zones provided around such power cables. During those times a famous American university studied the impact on plants and learned that corn, wheat and other crops grow better under electric power cables than elsewhere. They came to the logical conclusion that these electrical fields do not have a harmful impact on plants, but rather stimulate their growth. This scientifically proven and widely published 'big' result was once read by a farmer who burst out laughing. Corn and wheat grow better under the cables because the birds sitting on them drop a lot of excrement there and as a result, the field beneath is better fertilized! The scientific thesis soon failed.

What's the lesson from this? The blindness of rationality is that from the given information we come to a conclusion that seems to be the most logical one. If something is logical according to the given information, then it has to be true as well. We believe in it, build a whole system according to it, and are convinced about it. How many times have we fallen flat on our faces just because we came to a wrong conclusion in the absence of some particular information?

But who can have all the information? How much do we know about the world at all? We could ask this the other way around: how much we do *not* know? A big portion of purely rational decisions can only be wrong!

If this sounds ridiculous, please take a moment to think it over: a rational decision can be right only if all the factors are rational and known. Developing a car is like that. At the same time, the result of our rationally-built society and its rationality-based science is that we live in a quickly-dying world. There have never been as many unhappy people in this 'wonderful', rational world as there are now. If this all-powerful rationality were so great, would we have arrived here? How can it be that our incredibly professional rational world cannot provide any solutions to these serious problems? How can it be that this Holy Grail of rationality could not predict this whole situation? By means of rationality, our level of technology and convenience keeps growing, but we pay a huge price for it: the destruction of the world and boosting of our misery. When I started to see with my heart, I realized how different it is to live like that. I have realized many times since then how awkward and narrow-minded my purely rational decisions were. It is incredible how the belief in rationality can restrict people. It happens because when we become completely rational, we lock ourselves away from both our hearts and the Universe[6]. Our feelings, the wisdom of our mind, our creativity and our emotions are pushed into the background and suppressed, and as a result we get tunnel vision. We are convinced that we are making the right decisions. However, it is actually lucky if we make the right decision. If it's a purely rational question like solving a mathematical equation and we have all the knowledge and information needed to solve it, then our dry rationality brings us success. Moreover, in these cases it is the right decision to trust in it. But if it's about a person's physical or mental health or about our relationships, then it is only pure luck if rationality takes us in the right direction. With such a complex pack of problems which cannot be described mathematically, what are the chances that you have all the information and knowledge to make the right decision?

What is the solution? A reverse viewpoint can make it right! If our heart opens, it takes over the control of rational thinking. This definitely doesn't mean that we will no longer think in a rational way; it means

---

6   Universe can be replaced by God, Almighty or any other phrase matching your belief system

that our emotions, intuition and creativity are opened and become a part of our thinking. Our rational thinking broadens and we become much more objective. This is what I call heart-coherent objectivism. However, this does not mean that it would be right if our emotions ruled our thinking, which would send us too far in the other direction.

It's no coincidence that the level of consciousness of Reason is at level 400 where there is real objectivism. This is far from the dry world of rationality. This level is above Acceptance and below Love. Here the heart is already open and supports thinking. Einstein also lived at this level of consciousness. If you read his writings that were not about physics, it will soon be clear that he was a real philanthropist and an open-hearted person. The respect for Life that is reflected by most Nobel-prize researchers refers to the same thing.

To reach the level of real objectivism from the narrow-mindedness of a purely rational person can be carried out only with the help of the heart. I know this is surprising to a rational person, but someone who lives from their heart does not understand why this needs to be written down at all; it is all so obvious to them. But since I began to live it on a daily basis, the world has surprisingly cleared up for me. It is an even more astonishing experience to see in my surroundings the infinite series of a lot of wrong, purely rational human decisions which generate unhappiness and destruction. People ruin their own lives and those in their environment every day without allowing their heart to have a say in their decisions. Unfortunately, I also fall back to the old "style" on a daily basis. But the change has already started in me and it's wonderful to live this way.

If the researchers who published the thesis about electrical fields stimulating plants had listened to their hearts, they wouldn't have dared to publish their results. They would have felt that something was wrong. It would have galvanized their thinking, and their creativity would have led them to the mystery of bird excrement. Instead, they considered their results with their dry, narrow-minded rationality. This is the difference between dry rationality and heart-coherent objectivism.

I'm not saying that every purely rational decision is wrong; I'm saying that whether a purely rational decision is right or not often depends on sheer luck. Obviously, there are fields where purely rational decisions cannot be wrong. One plus one is two. Likewise, applying a long-proven engineering solution is usually the right, rational decision. But most decisions in life are not physical, mathematical or technical questions; more than half of our lifetime decisions cannot be made on a purely rational basis. Our spirit, heart, creativity, feelings, conscience, wisdom and hints from the Universe—for those who have not cut themselves off from the Universe—are all needed for these decisions. Perception opens due to the use of these and the world becomes wider; our decisions become more proper than the ones made by simple rationality.

If rationality was so infallible, we would have not led the Earth to the edge of dying out. Think about it: if Buddhist spiritual leaders led society according to Buddhist beliefs, how many more healthy spirits and how much more natural environment would there be on Earth? The Buddhists do not understand how we can be so blind, silly and narrow-minded. Conversely, we do not understand what they are talking about. We think from behind our over-rationalized hubris that they are out of their minds, but you don't need to be a Buddhist to understand what they're talking about. At the same time, I see why we are living the way we are. I have some bad news: we are living in a wrong and improper way! We are thinking in a wrong and improper way! If you don't believe me, just look around you. How many people do you know with a healthy spirit? How many healthy ecosystems are there in your environment? The bigger part of Earth's wildlife is ill and struggling! And we still know and believe, clinging on our rational thoughts, that we are heading in the right direction, and so we keep on building this social system, growing on its faulty base.

When our heart gives gentle direction to our thinking, then the **spirit and mind will be in parallel and real objectivism is developed, namely thinking steeped with wisdom.** At that point, unhappiness and the destruction of Life will cease. It is high time for Humankind to wake up and realize they have to change direction from the wrong path! Instead of the dry and pure belief system of rationality, our spiritual development has to become the most important aim of our lives again. **With the conscious raising of our level**

of consciousness, we can get closer and closer to the heart-coherent world of objectivism and in parallel our happiness will grow continuously and the destruction of our environment will be reduced. If you do this, I promise you will be happier and healthier year by year! What is more important than healing the people around you? It's worth the energy and stamina, isn't it?

2.14. What is happiness? Can we search for happiness consciously?

These questions are really profound, and everyone is interested in the answers. At the same time, we already know that the success of our individual search for happiness is one of the most efficient determiners of climate and environment protection. However, the meaning of happiness has become distorted recently, which is why it is crucial to make this question clear now to establish the meaning of happiness used later in this book. The meaning of happiness that became fashionable as a product of western civilization is Life-destructive.

First, let's deal with the second question: can we search for happiness consciously? The answer is obviously yes. As we've seen in the chapters dealing with the levels of consciousness, the things we desire are different at the different levels of consciousness. The higher the level of consciousness we reach, the less self- and Life-destructive the things we desire are, and the more self- and Life-supportive they are. When I was at low levels of consciousness I engaged in a lot of self-destructive activities, which were of course harmful to my environment. Addictions and games belonged to these activities.[7] (On low levels of consciousness it is desirable to be an alcoholic, food addict, extreme athlete, porn addict... the list is quite long). At low levels of consciousness it is desirable to have infinite power, to long for money, or to close ourselves into the prison of selfishness. At lower levels, hedonism is preferred. At higher

---

[7] Regarding games, I highly recommend Appendix 2 and Eric Berne's Games People Play

levels we feel the desire to help others and our environment, and to love ourselves. We would like to perform activities that build us and our environment; it disturbs us if we do or see Life-destructive things. The answer to the question may now be obvious to you as well: **it is not happiness that has to be searched for, but the way to raise your level of consciousness! This brings the more and more lasting happiness.** That is why spiritual leaders say that happiness comes from inside—not from chasing objects and services.

Unfortunately, the way in which you can achieve this cannot fit within the length of this book, but the next book will be about it. One thing is sure: continuous spiritual self-development is needed. It is as important as keeping our body fit every day and developing our knowledge continuously. If we had not attended school, we would be a "jerk", a nice way of saying mentally undeveloped. If we did not train our bodies on a daily basis, we would be thin or overweight depending on our personal features. Likewise, if we do not develop our spirits on a daily basis, we will remain at a certain level of consciousness and be stuck at our current relationship to happiness and the world. This is why a lot of people don't understand why they can't find lasting happiness however hard they search for it.

Now we can deal with the first question in the title of this chapter: what is happiness? We've established that happiness differs according to the different levels of consciousness, which is why it means different things to different people. One thing is certain: **the meaning of happiness changes continuously as you develop** (or regress). For example, when I lived at a more Life-destructive level, I always needed some sort of buzz. To keep busy, sometimes I desired to do something that was bad or not right—something "outrageous". These things made me happy at that time. Today, inner peace and being selfless make me happy. My meaning of happiness has changed as my level of consciousness has risen. What I thought about happiness was not right, although I was convinced that I thought the right way. Obviously, for that level of consciousness it was right. My activities caused a lot of struggling and environmental destruction indirectly, learning from which made me reach higher and higher levels.

Now I am grateful for that suffering because it motivated me to change.

Here's a final thought: is happiness equal to the feeling of happiness? My answer is definitely no. Addictions (like compulsive buying) bring short term happiness, but our spirit's huge void remains in place, because of which we chase newer and newer short-term feelings of happiness. It is an unconscious state which drives us even deeper spiritually into a more and more unhappy life. There is a saying that the road to hell is paved with good intentions. We chase these pieces of joy unconsciously with good intentions. Happiness is equal to the feeling of happiness if it is not temporary and there is no void afterwards, but it instead has the result of building our spirit. This kind of happiness makes your existence Life-supportive, due to which you will instinctively do a lot to aid Humankind's future! Not to mention that you will live your normal days as wonders! So please join the team of Life rescuers by investing part of your energy in your spiritual development. With these lines we have reached the end of the part of the book that covers all the important basic concepts and theoretical questions that were necessary for us to speak the same language. However, there have been consequences referring to the solution to climate change; the following chapters will focus on the introduction of the solution system.

CHAPTER 3

The solution: the Six Programs of Change

I consider climate change to be the biggest challenge in Humankind's history. It's a challenge that can either destroy the whole human race or raise us up into a long, peaceful and developed future. I believe in the latter! Everyone expects the politicians, scientists and governments to solve the problem, but if you've reached this point in the book you can clearly see that climate change is caused by about 8 billion people's small decisions, so it can be stopped by 8 billion people's tiny added activities! Now it is also clear to you that the fight against climate change and the growth in the average level of human happiness go hand in hand. At the same time, it is not our only task.

The reasons for climate change must be searched for in all of society and as a result, the problem is complex and diversified. To solve it, I developed a simple set of rules that I call the Six Programs of Change. With the help of the Six Programs, Humankind can reach a carbon neutral state! However, not only will it result in a neutral carbon balance, but we will attain harmony with Nature in every other field. Furthermore, on the individual level, most people will attain a more complete harmony with themselves and others than ever before. **Here we are standing on the threshold of the world peace that we have been wanting for thousands of years.**

We can reach Humankind's carbon neutral activities and at the same time be in harmony with Nature and other people with the following six programs:
- First Program of Change: The Program of Revitalization
- Second Program of Change: The Program of Agglomeration
- Third Program of Change: The Program of Population
- Fourth Program of Change: The Program of Happiness
- Fifth Program of Change: The Program of Society

- Sixth Program of Change: The Program of Economy

Every single program improves the relationship of People and Nature, others and themselves to almost the same extent. Of course, this is not the result of a precise tally, but rather a simple, approximate 'engineering' estimation that can show the importance of the Programs. The Six Programs of Change actually draw attention to the six main areas where transformation is needed and give the right directions for transitions and the ways of solutions. Of course, as I wrote at the beginning of the book it is only a framework that has the advantage of being easily adapted in any countries or regions to the local environmental, cultural and social systems. What is good in the system that it can be built up in a democratic way as a bottom-up initiative, but can be realized under political control as well. The other advantage of the system is that it gives an opportunity for change to those with power and wealth. This is important because if it were not so, change could only be started with revolution.

I will go into detail in connection with the individual's role in a separate, independent chapter after introducing the Six Programs, where we analyze what you can do within your life and I provide some practical advice. By the way, separate material about the practical opportunities is under preparation, and you can also always find an expanding range of information on our blog and website.

It is necessary to highlight that to discuss the Six Programs of Change one after the other makes them easy to understand. In reality, the Six Programs should be realized at the same time in parallel, as they aid and enhance each other.

It's a good feeling that particular initiatives of these Programs have already appeared in society, almost instinctively. This clearly shows that the change has already started on the tendency level, but we are at the very beginning and most of us have not realized it yet. So let's get started! Please join me on this exciting journey and discover the Six Programs that can save Humankind.

CHAPTER 4

The first program of change: The Program of Revitalization

4.1.1 Is really the whole world possessed by humans?

One of the biggest climate protection problems lies in the human approach. **Humankind thinks that it possesses the world.** Looking at satellite images, every piece of land is subdivided. Every area is owned by people, a state, a state union or organization. But we cannot possess the whole world—only a part of it! If we do not provide enough freedom to Nature, we will destroy it together with ourselves.

This is similar to the situation in medieval times when there was a good king and a bad king, but now it is on a bigger scale. The bad kings depleted the serfdom and the lower nobility to extremes in order to fuel their own luxury. As a result, the country declined. The end of these stories in history was either a revolution and the king was persecuted, or the weakened country was occupied by a foreign power. Good kings put burdens on their serfdom and their subjects as well, but their strict governance was never more than what the system could bear. These kings were usually loved for being fair. Justice and straightforward but strict rules were the basis for strong kingdoms. Humans have become the kings of Earth. Our power over the Earth is almost limitless, but if we go on in this way our power can only be temporary. However, if it continues to make us blind, we can only be bad kings. We weaken what supports us, and the consequence will be a revolution that will wipe Humankind off the face of Earth forever. This revolution has already started: you can read about its signs in the news every day, just as you could read about it in the beginning of this book.

The Program of Revitalization provides the realization of the king theory, in which we provide enough free living space to Nature to support us stably. We cannot rule over the Earth limitlessly or possess it as we like! **We have to limit our greed and desire for power!**

4.2. The protection of natural areas

David Attenborough obviously refers to this in his movie about climate change *Life on Earth*[1], in stating that **the fewer natural areas we have on Earth and the more biodiversity is reduced, the bigger the climate change is.** This is quite clear, as sooner or later carbon dioxide goes into the atmosphere from the decomposition of dead biomass. The film shows beautifully how we have destroyed almost two-thirds of our natural ecosystem since the 1950s. The research models have also proved that if we continue our way of living according to the current tendencies and the population grows at the same pace, the portion of the terrestrial natural areas on Earth will reduce to around 15% by 2050. The extent of the current destruction already suggests that Earth's ecological systems are standing on the threshold of a final collapse. One thing is sure: in 2050 the situation will be irreversible if we do not do something immediately!

It is therefore important not to classify as protected only those valuable natural areas which have rare species or other special living areas, but to protect *all* the areas which have a role in the climate system or in maintaining biodiversity on a global, continental or regional level. This definition means only one thing in the current situation: **every area that is still in a semi-natural condition must be protected immediately!** With this step we can stop the CO_2 emission surplus coming from environment destruction, and we can slow down the reduction of biodiversity.

---

[1] I recommend several films to you in a separate list at the end of the book if you want to spend your viewing time on valuable films about climate protection.

To sum up, from today on every semi-natural area must be considered valuable, and consequently must be classified as a strictly protected area, regardless of whether they are on sea, land or ice cap.

4.3. Main steps of the Program of Revitalization

The things contained in the previous chapter are important, but are not enough for success. To continue, let's examine the Program of Revitalization, which can be carried out in three main steps:

Step I: stopping the destruction of every semi-natural area that already exists.

Step II: every area that is degraded slightly must be recovered by human effort. (This is what we call revitalization.)

Step III: the natural and revitalized areas must be expanded until human emissions and the ability of Nature to bind pollutants are in equilibrium. (Not only regarding GHG, although it is the most urgent.)

It's good to see and understand clearly that the expansion of natural areas can extract a huge amount of CO_2 from the atmosphere and consequently slow down climate change and cool down the planet gradually. It is also important to point out that in the areas which should be revitalized, there may never be the same ecosystem as what was originally there. However, these renewable habitats can support the recovery of our ecological systems properly, even if they will not be perfectly as they once were.

It all sounds great, but perhaps you're instinctively asking the question: how?

For Step I, a Natural Area Registry must be made. All the nearly-untouched areas have to be recorded in the registry. If every region does this and enters it into a common database, soon a Global Natural Registry will be developed. Afterwards, all the nations must offer them to the global climate interests and provide their protection.

To understand the program, it is crucial to see that it is not only about land areas, but also the areas covered by water or ice, including the ice caps and oceans. It is particularly important to know that the role of oceans in binding CO_2 is much bigger than the ecosystems on the land areas, so we have to take it seriously! At the same time, we mustn't wait with the Global Natural Registry for a global agreement; the cities, regions and nations have to start to join to this system, so it will gradually become global. Sooner or later, everybody will admit that this is the right way, and there will be significant social pressure on the ones joining later. It is also easy to admit that without it we will die, so the need to act will occur everywhere on Earth! We do not have to wait for international agreements; they will just happen anyway if efficient initiatives come from several places.

4.4. Rules in connection with natural areas, Natural Territorial Development Directive, revitalization

After the natural areas are taken into protection, every kind of human activity on them will be forbidden. Exceptions are revitalization activities, searches, and educational human activities. How can we avoid illegal logging, forests being burned, and several other Life-destructive activities in these areas? I promise the Six Programs will give you the answer!

While carrying out Stage I, a Natural Territorial Development Directive must be created, which has to be accepted locally by the initiators. If there are a lot of such directives, then it is advisable to unify them in an international agreement, but this is a further step. From this directive a Global Natural Territorial Development Directive will gradually develop, an agenda that will decide which areas are considered to be slightly damaged. Its basic concept will be to determine where the economical limit of recovery is, and which areas can be restored with a relatively small budget. The principles of cost calculation and setting out areas must be fixed in this directive.

Afterwards, the areas needing restoration can be set out and revitalization processes can be started. It is already a proven fact that these processes can be conducted without economic disadvantages. It is also proven that this approach is more economically viable than restoring the increasing damage caused by climate change—and more effective as well (Paul Hawken 2020).

It is crucial to point out that it's necessary not to wait until the whole world accepts this. Some countries have to start it, then more and more other countries will slowly join. It will grow slowly, as it is common interest that brings countries together.

A significant aspect is to connect the natural areas into a network as coherent as possible in order to reduce the fragmentation of habitat and vegetation. This is crucial for enhancing the assimilation capability of Nature and to control the reduction of biodiversity[2]. Habitat and vegetation fragmentation is one of the most serious causes of climate change and biodiversity reduction. The Program of Revitalization is not enough in itself, so it is necessary to put it on a system level. The solution is disclosed in the following Program.

While the second development stage is going (over more decades), the developments of the other five Programs are running in parallel, as a result of which the human emission of pollutants will first stagnate, then become reduced. At the beginning of Stage 3 of the Program of Revitalization, we have to take into consideration the assimilation capabilities of the Earth's natural areas and the actual extent of human emissions. From this result, it can be calculated how much territory is still needed to give back to Nature to bring the impact of human activities and Nature into equilibrium again. If this is disclosed, the process can be carried out according to a plan, systematically. This is the most expensive step, as in the first two stages we already restored and provided the protection of the areas still in good condition or that can be restored easily.

---

[2] assimilation means the binding and absorption capabilities of natural resources

4.5. Providing the protection of natural areas

How can the protection of natural areas be provided? The first step is that **Nature must be given rights.** Accept that every living creature has the same right to Life as you. This is the basic principle of the science of ecology, and at the same time it delivers the principles of Buddhism. They believe that when you cut down a tree, that tree could be the current manifestation of your dead grandfather. Your grandfather would definitely sacrifice his life to you as he loves you, but you would accept it only if it is absolutely necessary! In other words, you can cut down trees only if it is highly important in terms of your life. If you respect Nature and accept that its right to Life is as important as yours, then you don't take a shortcut between two sidewalks and step on the grass unless you are really in a hurry. This complies with the basic principle of the science of ecology that states that **every species is equal.** The institutionalization of these principles means that we give rights to Nature. Without this, future society cannot exist, since this is the only way to build cooperation between Humankind and Nature.

People have always believed that the Earth is exclusively theirs and that they can do what they want with it. We have to realize it's not true! The more freedom we give to Nature, the better our lives will be, so the time to make a change is now! After we set aside the areas of the Global Natural Area Registry, Nature will have primary rights on those areas. Anyone who acts against Nature can be condemned and obliged to restore the area or pay compensation. The practical implementation would require setting up an organization that represents Nature's legal rights, with representation in every natural area. These organizations have to be given the necessary authority to represent Nature legally. From that point on, if someone is seen littering in the woods or cutting down trees illegally, they can be sued in the name of Nature and the local organization can take legal steps for its protection. There are already such organizations in different countries, but unfortunately they have little power, few legal tools, and insufficient money to play an active role. At the same time, it is important to point out that these organizations will be needed in the decades after launching the Program of Revitalization.

Due to human spiritual development and the increasing respect for Nature, less and less Life-destructive activity will appear in society. With the increasing acceptance of each other and Nature, Humankind will transform into a species with a caring, confidential relationship with Nature. In the distant future there will be no need to talk about sustainability, as it will be fundamental.

The next step will be to set up the local **Nature Conversation Guard.** This organization can take action only against illegal environment destruction, and its responsibilities will be the coordination and socialization of local revitalization activities and to provide the necessary resources. Once there are such organizations in many countries, a global one could be set up. Every member country would delegate resources and tools, and provide funding. The task of this organization would be to control the protected natural areas, prevent any illegal environment destruction, help the revitalization of natural areas, eliminate forest fires, provide resources to tree planting programs, etc. This organization has to dedicate all of its efforts to serve the fight against climate change. The global organization differs from the local ones in that it provides the protection of the wide areas under international protection and oversees their revitalization processes. Furthermore, it helps the areas experiencing extreme hardship after natural disasters.

In parallel with this step, a **retraining and economic support** system must be developed with the aim to provide a living for those previously earning a living from the destruction of Nature (like loggers) and transferring these people into other sectors. How will the funds be provided for this? A solution is offered in the Programs later on, as the Six Programs of Change support and enhance each other. By the end of the section you will have the big picture about this system.

4.6. Social rudiments

If you think about the principles of the Program of Revitalization, you can see that **the rudiments of the processes introduced here can already be found in society.**

Greater numbers of nature preserves are being set aside worldwide, and there is an increasing public-social demand for it. There are more and more public demonstrations because of the destruction of meadows or wetlands. Local people will sometimes even fight to save a row of trees—we have the inner instinct to do that. However, it is a better idea to remove trees in the city center and plant them instead in other places where they can support an ecological system. I'm certainly not advocating cutting down all the trees in a city center; I'm advocating using our limited energy spent on environment awareness systematically to maximize sustainability and effectiveness.

There is a legal regulation in Europe on ecological networks which aims to mitigate the impacts of habitat and vegetation fragmentation. Unfortunately, this regulation is not enough without the Program of Revitalization and Agglomeration, but it clearly shows there is social demand for the interventions.

At the same time, the EU's Water Framework Directive has existed since 2000, and is the same as the Program of Revitalization but confined to European waters. Its positive effects were clearly evident by 2021; the Water Framework Directive has reversed the deteriorating tendency of European waters. The basic concept of this directive is that every water that is in natural condition has to be preserved in that condition, and all of the waters that can be brought back to a semi-natural condition have to be restored to that level. If it works in the case of every body of water in Europe, why couldn't it work globally for every element of the environment? The social system of cross-border legal regulation already works, so the process has already begun to bud. Nothing else is needed to attract politicians' attention besides a lot of people's honest determination.

A further good example is that the UN is currently working on placing 5% of ocean areas under protection.

This is an extremely small portion, but a good beginning!

I have read about a Brazilian family who planted forests on their huge land areas, which are totally natural, diverse rainforests today.

I also know about an initiative in Indonesia wherein every school graduate has to plant 10 trees. The Indian Tree Lovers initiative was also successful, with several large areas being reforested.

There is huge power in social collaboration! The Program of Revitalization is nothing other than the growing of these sprouting seeds into a bush until it becomes the Program of Global Revitalization.

4.7. The benefits of the Program of Revitalization

To understand the changes generated by the program, let's look at an example through the issue of wood. It is important to point out that this example helps the better understanding of the process and can be transferred to any other natural resources. If we decide globally (or gradually by having even more countries join the system) that no more natural forests can be destroyed from now on, then the consequence in society will be a wood shortage. In order for this not to happen so suddenly and dramatically, a deadline must be given for when the rule enters into force. Say for example that it comes into being in 15 years, and until then we are going to gradually reduce the natural extraction areas. In the meantime, we provide subsidies, tax allowances and other support to accomplish artificial forest plantations and also help the continuity of people's incomes through re-training. For example, it is practical to plant Oxytrees or similar species, since such trees are able to bind double the amount of CO_2 from the atmosphere in a given time as a traditional tree. There are more than 30 types of such hybrids. They are not invasive, and if you cut them off at their trunks they grow again, therefore you do not need to plant new saplings. The wood is high quality and can be used for everything from fuel to timber. That being said, heating with wood is obsolete today with the technical knowledge we have, and we are slowly reaching the point where we only burn wood for its old-fashioned ambience.

It is important to point out that artificial woods do not represent such a high ecological value as natural forests. The forest plantations that provide wood must be located outside natural areas. Through the revitalization of natural areas, the aim is to help the re-settlement of organism groups. This must not be confused with the issue of artificial forest plantations.

If we make such a provision and support artificial forest plantations, Humankind will be able to change in such a way so as not to cut down a single tree in natural or semi-natural forests. This change will have social, economic and environment protection consequences as well. Let's look at them one at a time.

Due to the change, the price of wood will go up. This will have a lot of positive consequences in society, as we will not waste wood as much as we do now. Think about how much furniture we throw away just because it isn't fashionable any more. And also take into consideration that the products of the actual furniture industry are not about durability at all. The furniture made by my grandfather 100 years ago is still in use. Conversely, I have already taken apart and assembled my bed at home three times but I cannot do it for the fourth time because it is going to fall apart, even though it is only six years old. It is only design that matters now; durability has not been a factor for a long time. If the price of wood goes up, the price of furniture will go up as well. If this happens, we will consider durability first when purchasing furniture. Furthermore, the social recycling of wood will be much more effective. If something is expensive, it is worth recycling. If so, the demand for raw material will be lower and the production of artificial forest plantations and social needs will therefore approach a middle ground. If wood is expensive, it won't be used for heating; it will be more worthwhile changing to renewable sources for heating and cooking. The whole process would change from a downward spiral into an upward spiral. The furniture industry will be transformed as well. Mass production will be reduced, resulting in job loss and an economic decline of some areas of the sector. On the other hand, furniture repair services will be revived, creating new job opportunities, and even more jobs will be created through the demand for forest plantations. In this way, people's livings will not be threatened—only the allocation among the sectors in the wood and related industries will be restructured. The gradual change will make it possible to accomplish this without economic losses.

What will change in a positive direction? So far, we humans have been able to go into the woods and been free to destroy them. Obviously, the legal regulation regarding this issue varies between countries, but compared to the principles of the Program of Revitalization it is definitely the case. Forest destruction has accelerated as never before. In the Amazon forests alone, an area the size as Hungary is cut down every year (this is not only about wood, but it is just an example). Forest destruction will stop with this intervention. With artificial forest plantation we can increase the extraction of CO_2 from the atmosphere more than ever. The trends will finally reverse! We are going to create Life instead of destruction. We will start to live in a more Life-supportive way. We will save the remaining biodiversity bases on Earth and give them a chance to strengthen again.

If we put the proposed forest plantations in particular areas, they can provide transition zones between areas intensively used by people and natural areas, thereby enhancing the tranquility and protection of natural areas. These are wonderful changes, aren't they? Not to mention the fact that the recovering soil structure under the trees of a forest plantation can extract more CO_2 from the atmosphere than trees! Nothing else is needed—just to change the human approach and political power to consider this issue as their target.

I am convinced the social support will be big and will grow in the future. The current increase in community tree planting programs shows that the social need exists and the process has begun. What is important is that whatever we do should be done systematically and efficiently at the program level.

Through the example of wood, we can see that the implementation of the Program of Revitalization does not cause serious damage to society or the economy—only restructuring. With progressive phasing in and proper attention from the government side, people and companies can be led through the changes easily. This process can work with all the natural resources that are provided for us by Nature for free, as unbelievable a gift as that seems.

If we implement the Program of Revitalization, what positive impacts it will have on our society? I've summed them up briefly here:

I. The reduction of biodiversity will stop, the extinction of species and soil degradation will slow down, and Nature will gradually regain strength.

II. **We will extract more CO_2 from the atmosphere than we pump into it**, therefore climate change will slow down and we will slowly cool down the planet again.

III. The price of natural resources will increase, so over-consumption will be reduced and recycling rates will grow.

IV. People will respect Nature again and consider natural resources as values again.

V. The identification with Nature and the respect for Nature will have a positive impact on our spirits, therefore human society will be more peaceful and harmonious. There will be less individual and social tension.

VI. We will prevent the limitless expansion of Humankind on Earth.

VII. **People will head in the direction of being in balance with Nature.**

What can you do to aid all of these? A lot! A separate chapter will deal with this at the end of the book, but first I would like to show you the big picture of the whole system.

CHAPTER 5

The second program of change: the Program of Agglomeration

National selfishness is a serious barrier in the fight against climate change. We can clearly see how limited the results achieved by climate conferences have been, despite being held for decades. The reason is that every nation-state considers those agreements according to their own narrow-minded and selfish interests. We live in a world of international competition where nobody wants to fall behind and everybody wants to get more from Earth's cake. This is the global manifestation of the notion of a nation-state. The Program of Agglomeration and the Program of Society offer a solution to this. Selfish political leaders are elected by selfish people. As a nation's average level of happiness rises through its spiritual development, the more selfless and conscious leaders we will choose. It will also help to accelerate the processes of change.

5.1. Targets, basic principles

The target of the Agglomeration Program is to divide the entire Earth into agglomerations and natural areas. We have already discussed the meaning of a natural area, so it's time to check what we mean by an agglomeration.

An agglomeration is a demarcated area consisting of one or more settlements, which can be surrounded only by a coherent natural area. The previous Program mentioned that Nature has to be given rights. The Program of Agglomeration completes the Program of Revitalization in that in natural areas Nature has primary rights, while in the agglomerations People do. Therefore, the Nature-People balance is fulfilled as the area and rules of operation of Nature and People become separable both in a legal and a territorial sense. It is crucial to see that the final aim is not the final separation of Nature and People, but to provide space for the strangled natural systems to recover.

The demarcation of the agglomerations is independent from countries. They have to be determined according to the natural conditions affecting the limits of the agglomerations, but of course during the transition time everything happens with respect to all countries. This is to say that in the case of cross-border agglomerations, the countries collaborate in developing the agglomerations at the beginning of the process, just as it works in the case of waters in the Water Framework Directive.

First, we have to take stock of every natural or semi-natural area (floodplain, mountain forest etc.) whose basis should be the natural areas that are already under protection. These areas must be connected in a way that the natural areas create a cohesive system. To reach it, the agglomeration boundary formation rule is that the boundary lines of two agglomerations cannot touch each other, and there must be at least 1-3 kilometers between the boundary lines of any two agglomerations. (The minimum distance must be decided by ecologists taking the local ecological system into consideration.) The agglomerations should be imagined as separated islands within the cohesive natural areas. The natural areas are divided among the agglomerations considering the sizes of each agglomeration's assimilation capability.

The base of the balance between Nature and Humankind is given by the agglomeration model. The basic rule of the agglomeration is that every agglomeration has to develop and establish its social, technical and economic systems in a way that the agglomeration's emissions should be in balance with the assimilation capabilities of the surrounding natural areas. Emission does not only mean the emission of GHG, but any kind of pollutant that gets into the air, water or ground. It is important to highlight that in calculations and counting, long-term processes have to be taken into consideration, as the time scales of changes caused by people and the changes happening in ecological systems are very different. From a technical and scientific standpoint,

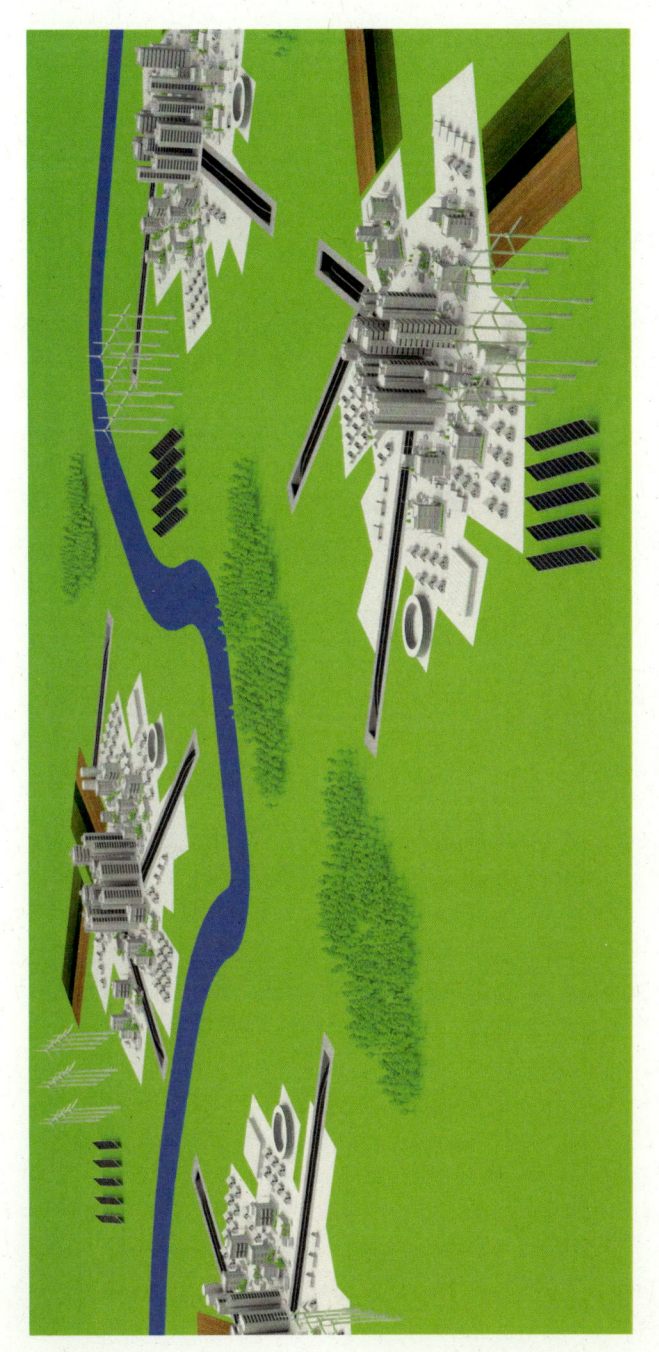

Figure 4: Agglomerations and natural areas

Even though the drawing is schematic, it illustrates the relationship between agglomerations and natural areas. You can see that the agglomerations are located as islands in the almost-cohesive natural area. Longer sections of the roads going from one agglomeration to another run underground so as not to break the cohesion of natural areas. Of course, the agglomerations can be very different in size and character, so you should not be misled by this idealized drawing. An agglomeration can be a village or a modern metropolis. The point is that every human activity occurs within the agglomeration's territory.

we are already able to operate such agglomerations; nothing else is needed but to set our thinking in the right direction and to carry out the necessary social and economic changes. I am going to go into details while introducing the other Programs.

Within the agglomerations, every human activity must be solved: agriculture, industry, trade, recreation etc. because, as was mentioned in the Program of Revitalization, no human activities can be performed in natural areas except for some specific activities.

The agglomerations are linked by roads and utilities of course, but not in the way that settlements are linked today; the agglomeration pairs can be linked only by one high performance transport and utility lane. Its aim is to avoid habitat and vegetation fragmentation as much as possible since the bigger that fragmentation is, the less biodiversity there is. This is the basic ecological rule discussed in the previous Program. The high-performance transport and utility lanes have to be established in a way to cause as little disturbance to natural areas as possible. Highways for example must be modified to have more wildlife corridors and tunnels on them than they currently do.

Regarding utilities, every agglomeration has to strive to be self-sufficient. The electrical grid is an exception. All the agglomerations must be electrically interconnected, establishing a Global Grid[1]. If total self-sufficiency cannot be solved somewhere in terms of utilities except for electricity and the IT network, in these places a connection between two or more agglomerations can be allowed, but these pipeline connections must be implemented on the same paths as transport connections between the agglomerations. However, as a basic concept we have to strive to be self-sufficient in order to be able to develop eco-efficient agglomerations.

Organizations that protect Nature's rights must be set up in every agglomeration, and they must be given appropriate rights so that every agglomeration will take care of the protection of the surrounding natural areas. This organization will conduct the monitoring[2] of the natural areas

---

1   Grid means network
2   This means to check the ecological systems' condition with measurement

and govern the necessary environment protection and revitalization activities.

The method for establishing agglomerations must be developed in detail in a document called the **Agglomeration Establishment Strategy** in order that the establishment of agglomerations happens properly in every respect. Agglomerations can be set up as a result of a global agreement, but it is more realistic if they emerge organically as the leaders of the settlements realize that this is the only chance to save their future. In this way, more and more settlements will organize into agglomerations, and more and more agglomerations will be established. Certainly, the central regulations can accelerate the process. Upon the establishment of an agglomeration, the Agglomeration Code of Ethics comes into force, setting the main orientation of the society in a constitution-like way, and of course it has to be put together precisely. In this book I am going to introduce only the main principles of the **Agglomeration Code of Ethics:**

1. The agglomeration strives to reach and maintain the perfect balance with the surrounding natural areas.
2. The main aim of the agglomeration is to reach the highest possible level of global happiness.
3. The agglomeration selflessly takes part in conducting the global tasks necessary in order to realize the greatest happiness and balance with Nature globally as soon as possible.
4. The agglomeration considers global aims ahead of its own local aims. In agglomerations, the principle 'think globally, act locally' has to be applied. An exception is the protection of local values (e.g., culture).

5.2. The principle of interdependence

Human society owes its development to collaboration, and its specialization is built on collaboration. Social collaboration means that we trust each other and in each other's special knowledge. Just imagine if the electricity supply stopped overnight on Earth. In about a half a year, more than the 90% of the population would die out in the more developed half of the world. If we couldn't trust that the electricity supply would be continuous and stable in the future, this single factor would be enough to break the development of Humankind. And this is the case with everything. We trust there will be someone to repair our cars or produce new computers or provide running water from our taps. We clearly have to trust each other! The social system today solves this statement with the strengthening of self-interest. It increases everybody's interest in order to that this self-interest gives rise to a small segment of society. Everybody makes money. For this money, they have to undertake some kind of job. And work generates the added value of society and therefore its sustainability and/or development. Everybody is a little gear in a big machine. This is admittedly an extremely simplified model. Here, everybody's self-interest is forced to be part of the system, which is not able to deal with the more complex needs of Man and Nature.

It would be difficult to imagine a simpler social system than this. It could spread due to its simplicity and because the global level of consciousness has not made Humankind capable yet of a higher level of social system. Capitalism is not bad in itself; it was an important stepping stone in our development, but it is old fashioned today, and must be followed by a more exciting and human-oriented system.

In order to understand what a higher level of social system would be like, let's take a look at how Nature and the human body function. The cells in the human body are specialized as well, but every cell is a part of a bigger system. Every cell has self-interest although its role in the big system takes a higher priority. If the cells do not perform their tasks according to the interest of the big system, they will die. This is how the wonderful organization called Humankind will be. Nature functions in the same way. There is some self-interest at the individual level, but in the system of Nature everybody depends on everybody. If any species that is part of the system dies out, it causes the torsion and degradation of the whole system. The specialties of living creatures can be imagined only as a perfectly harmonious element within Nature's system. In any case, the species dies out as the evolutional backlog reaches it either way. You can clearly see this, as we have already discussed that competition can only be a lower-level organizing principle. Collaboration is the main driving force of evolution.

The human body and Nature's systems work according to the principle of interdependence, as these are the rules of efficient collaboration. To live according to this principle results in a much more efficient and higher level of social system than the current social model, which feeds infinite selfishness.

The condition for human society's further development is to move on to this level. We have to develop from the capitalist system—which encourages infinite hedonism—in the direction of a society based on the principle of interdependence, and to do this gradually. Today, the social transition is possible within two or three generations. Humankind is ripe for it, but most of us simply don't think of doing things differently than we are used to. The fact is that a well-functioning, healthy and happy family lives according to this principle—mostly unconsciously. The family unit stands above the family members' self-interests. Of course, the family members have their own self-interests, but these must be often contained for the good of the family. Self-interest can expand until it challenges the family's interests. Everyone trusts everyone, so common happiness is able to rise due to the interdependence.

I always try to implement this principle in my companies as well. When I was a selfish leader, my companies were far less successful than they have been since I began trying to implement the principle of interdependence. I started three years ago. The development processes have sped up in my companies since then, and my colleagues feel much better! The employee turnover is approaching zero. The model is so good that I am definitely going to continue in this way. It is not by coincidence that in modern leadership training and in the literature dealing with the conditions for success the principle of interdependence is argued as being a cornerstone.

If it works at a family or company level, why couldn't it work socially? Actually, deep in their hearts most people hate being selfish, and this incredibly selfish society torments most of them spiritually. Almost everybody complains about how they suffer from other people's selfish and stubborn attitudes. The need for interdependence is clearly part of us. Let's talk about what it has to do with the Program of Agglomeration.

Future society will be entirely global, where agglomerations will be the cornerstones of global society. They will work like the cells of the body or as members of a family. It would be akin to the alliance of city-states of Ancient Greece. The system will be permeated by the principle of interdependence. Every agglomeration has self-interest, but neither of the agglomerations' self-interest can enjoy an advantage over global targets. Every agglomeration acts locally but while taking global interests into consideration. The global target is to attain the highest level of happiness and to reach the greatest balance with Nature. Keeping the same direction and with common effort we will be much more efficient than now, when we're competing at each other's expense at the individual, company, settlement and national levels. There won't be competition between the agglomerations as everybody takes responsibility for their own territory regarding the balance of natural areas. At the same time, the mission system (to be detailed later) will be built on the supportive attitude of agglomerations instead of on competition.

Why does this result in a much more efficient and happier world than the current one? Today, nations' self-interest drives the world into destruction. Every country is a sovereign social system which competes with the others. Every country considers only its own self-interest and they don't want to let up, so international agreements are possible only through dramatic compromises if they are possible at all. Change can only be slow and development distorted. I have already mentioned that Brazil cuts down a Hungary-sized rainforest every year. We know that the further destruction of Amazon rain forests generates the crossing of a balance point which causes the destruction of the global climate system, but we can't do anything about it. Why? Because the short-term self-interest of Brazil goes against it. This is the way that nations' self-interest destroys the Earth. It is the competition among nations and infinite greed that drive the world into death. The competition is the same between individuals, cities, companies etc., so it's clearly time that we move up a level.

We should accept that we People depend on each other and on Nature. We should admit that interdependence is the only right way, instead of the destructive social system of selfishness. The change to the principle of interdependence is crucial for individuals, companies, settlements and countries in order for us all to be able to move to a higher level of development.

5.3. Global Grid

The Global Grid is the practical realization of the principle of interdependence which speeds up the development and spread of green energy systems. As already mentioned, the agglomerations each have to strive to solve their utilities needs within their own territory, except for electricity and the information network. District heating, fresh water supply and waste water disposal etc. are utilities to be solved on the spot. When setting the size and borders of agglomerations, these have to be considered. If it cannot be solved, it is of course acceptable to receive healthy drinking water from another agglomeration, but this can be permitted only if it cannot be solved otherwise. By contrast, I imagine the electricity supply very differently.

Every agglomeration joins the so-called Global Grid, a worldwide electricity network which goes through every agglomeration. The Global Grid works on the principle of interdependence. Every agglomeration has to keep two rules in mind:
- we have to move in the direction of the biggest ratio of electrification that is possible within the agglomeration. (For example, replacing gasoline-powered cars with electric cars,

changing gas heating to electric heating, etc.)
- the agglomeration has to strive to change to the biggest amount of renewable energy production

What will the result be? The ratio of renewable energy will continue to grow within the energy mix and it will become possible to stop using fossil fuels almost completely. This ideal condition can be reached without building redundant capacity, so globally we can reach it the fastest way. The smallest possible energy storage capacity must be built, considering its global amount. The remaining emissions of fossil fuel can be bound by CCU/CCS technologies.

Why is it so obvious? The great disadvantages of green energy systems are that their availability is barely controllable. We cannot adjust our energy needs to when and how much sunshine there is or how strong the wind is blowing. This is why the larger the proportion of green energy systems is in the energy mix, the more unstable the system is or the more energy storage capacity has to be built. But the latter is really expensive and has not yet been solved economically. (However, exciting technical developments are being launched regarding this topic on a regular basis.) At the same time, this problem is mainly solved by the Global Grid, since the surplus of some regions' green energy production can be used to satisfy other regions' energy needs. If we look into the topic from the aspect of solar energy (obviously the situation will be simpler due to there being several green energy systems), on the part of Earth where there is night, the amount of electricity can be provided by the Global Grid which is being produced redundantly on the other sunny part of the Earth. Imagine the Global Grid as being like the World Wide Web, and it becomes clear that the solving of seasonal, daily and other inequalities is realistic.

Obviously, there are at least two technical counterarguments that come to mind. One is the disadvantages of transporting electricity over great distances, and the other is that installing large cables under the oceans is not a realistic way to share an electricity supply among continents. Some experts may think of how green energy systems are affected by weather, which enhances the uncertainty of the system.

To technical questions, technical answers should be given. A New Zealand-based startup has developed an invention—attributed to Nikola Tesla—with which electricity can be transported over long distances with little loss and without any cables (EMROD 2021). So the transmission of electricity among continents seems to be solved with the help of today's modern technology. At the same time, our global weather forecast technology is already so accurate that a reliable Global Grid could be made to function exclusively with green energy producing processes. Obviously, fossil-based or nuclear power plants can be kept as emergency reserves where there is a realistic need. Nothing else is needed but to accept the principle of interdependence, as in: I give electricity in daytime, while I get at nighttime. Everybody depends on everybody, so nobody has to fear anybody. Furthermore, electricity would be cheaper than ever globally in this system. As a matter of fact, using this system we would not have to pay for the electricity produced by green energy systems; it would be free to everyone. We have to pay for two things: the transport of electrical energy and the energy coming from fossil fuel or nuclear-based energy production. However, everybody has to build enough green energy-producing capacity to be able to overproduce periodically, to meet their obligation to deliver into the Global Grid. As there are only a few among the green energy systems that can produce continuously and stably and the consumption always changes as well, most of the agglomerations sometimes deliver energy, and sometimes they receive energy. They do not pay for delivery but for the receiving, as in this case energy is transported to them. This can finance the system. The further the energy is transported from, the higher its cost. One part of the income will be spent on maintenance, the other on the development of the Global Grid, resulting in the gradual decrease of energy price on a global scale. If we do not have to pay for delivering energy, it will reduce the operating cost of our electrical system, since if the Global Grid uses our overproduction, certain power plants should not be set back, which has huge costs.

Now you can rightly say that it will not work, as in a system like this every agglomeration will strive to increase its energy storing capacity in order to avoid paying for energy coming from far away. The fact is that energy storage is more expensive than receiving green energy, and energy storage can be taxed if it does not serve global interests. Selfish energy storage is therefore possible within the Global Grid but unnecessarily expensive. The professional organization that operates the Global Grid has to establish energy storage capacities to provide a stable energy supply, but in such places and hubs with such capacities will be able to maintain the stability of the Global Grid. These energy storage locations do not belong to agglomerations but they are parts of the Global Grid system. The Global Grid is supported by the agglomerations, so the principle of interdependence is applied at this level as well.

The process described above can be actually found in its traces. The electrical energy networks are more and more intercontinental and the electricity providers provide in bigger and bigger areas. The increasing ratio of green energy systems forces the system to have more consumers in the common system because this leads to the probability of balancing the system's growth while the need to operate energy storage and fossil-based power plants will reduce.

The barrier to establishing the Global Grid is in the way in which nations think about energy policy. Every country wants to build energy overcapacity for three reasons: one is to protect its own energy supply, the second is that energy export is good for business, and the third is that cheap energy is one of the basic parameters of global economic competition. So globally, the self-interest of countries will result in building an unrealistic amount of overcapacity. If we think at the country level, increasing the ratio of green energy systems goes with the increasing of electricity price because of the increasing regulatory costs of the system. More and more countries have to face with the problem nowadays where green energy plays a bigger role in the energy mix due to the developments of climate protection.

In the case of the Global Grid there is no need for overcapacity; energy is cheap for everyone. Nothing more than courage is needed to step out of the world of national selfishness and into the more successful world of interdependence. Now we clearly understand why.

5.4. Balance strategies

Imagine a future where agglomerations work stably. The world seems to have agglomerations surrounded by wonderful natural environments. In the sea of cohesive natural areas, the agglomerations appear as islands. Nature and Humankind live in balance with each other. This is the basis for long-term development and the flourishing of Humankind so that our culture can survive and our descendants can live happily. In this future, all the pollutant emission of agglomerations is not more than the assimilation capability of the surrounding natural area. Try to imagine a world like this. It's a wonderful picture is, isn't it?

However incredible or utopian this dream seems, I know we can reach it in three to five generations. We are already prepared in the scientific and technical sense; the only question is how fast the social-economic transition will happen. Nothing else is needed other than a novel way of thinking.

Now let's start with a current rudimentary agglomeration. Suppose that the rules about setting up an agglomeration are accepted by the EU regarding their own territories, and the organizational process starts. Accordingly, some settlements are gathered to form an agglomeration. The current picture of agglomeration today: a few natural areas found in patches, a road network crossing every stretch of land, and a lot of agricultural areas and scattered settlements in the agglomerations. The agglomeration's emissions are many times more than the assimilation capability of what remains of the surrounding natural environment. The system is in a total imbalance. How will it transform into the ideal situation that I've described?

To understand this, I would like to introduce the balance strategies of agglomerations. The point of these strategies is to determine in what ways there can be balance between the agglomeration and its surrounding natural environment in the medium term. Here's a list of the balance strategies an agglomeration can employ, which will be written about in detail later on:

1. Reducing the population (e.g., via resettlement) in a democratic way.
2. Reducing certain industrial emissions by reducing production or relocating the industry.
3. Building highly efficient energy and material management systems.
4. Using low emission technologies.
5. Developing highly efficient cleaning systems for problematic emissions.
6. Altering local consumption habits.
7. Building emission-binding technologies (e.g., CCU, CCS).
8. Reducing its territory and increasing the natural areas around it.
9. Increasing the assimilation capabilities of the existing natural areas through revitalization.

Let's look at these strategies one by one.

Reducing population: this can happen only in a democratic way of course (as with everything else in this book). Populations are falling naturally in certain regions. Some politicians these days experience this as the end of the world, but this is not the case, as will be discussed in detail in the Program of Population. The point is that the reducing of a population is not a sin and is not bad! At the agglomeration level, migration can be limited with local regulation, and moving away can be supported with different incentives. Reaching a balanced population in an agglomeration can be gradually achieved with these tools in three to five generations. This topic will be supported effectively by the three methods introduced in the Program of Population.

Reducing industrial production: industrial production could be so high in certain agglomerations that the resulting emissions cannot be assimilated by the surrounding natural area—not even with cutting-edge technologies. In these cases, the agglomeration can make a strategic decision on reducing local production. This does not mean to reduce the production volume, since no economic operator would accept it. It means the relocating of certain production processes to other, less loaded agglomerations. This principle plays a huge role in getting the poorer regions caught up, and in the expansion of economic prosperity. Where poverty is greater, the emissions are generally much lower, so factories can expand in these directions. For example, Africa is still the only still carbon-negative continent.

Building highly energy efficient and energy saving systems, enhancing material management: the agglomeration is ready to do everything in favor of reducing energy consumption regarding the consumption of either its industrial, domestic, transport or other sectors. The target is to minimize its energy consumption with the best technologies and to reduce its emissions along with it. For instance, only LED lights can be used in the agglomeration, or every internal combustion engine is changed for an electric engine. The energy efficiency of electric engines is more than two times higher than for other types. Actually, these developments are the most popular ones today as well. The continuous development of material management reduces the raw material demand and the impact on Nature in several ways, so the actions of an agglomeration towards balance can be enhanced in that way too.

Building out emission-negative technologies: this strategic pack contains every intervention which can either mitigate or stop emissions from certain activities. Renewable energy processes, regenerative agriculture and electric transport belong here. Of course, in the case of some concrete measures there can be some overlap between the previous strategy and this one. The point is to reduce any emissions as much as possible, due to which the demand for natural areas needed for balance will be reduced.

Changing local consumer habits: an incredible amount of emission reduction can be reached by changing consumer habits. The agglomeration can launch a targeted information campaign to support the case or charge the environment-destructive habits with an extra tax. Alcoholism for example dropped dramatically in Finland when alcoholic drinks were charged with one of the highest taxes in Europe. Why couldn't the same be done with beef, palm oil, milk, coffee, or ice cream? All of these cause far more GHG emissions than any other food. I do not use any of them and I am not less happy at all. (All right: I drink 1 or 2 coffees a month).

Building highly efficient cleaning systems or emissions binding systems: this strategy targets the extraction of emissions before they reach the natural area's border. These can be accomplished by cleaning equipment installed on factory chimneys or wastewater treatment plants at the agglomeration's border. The agglomeration can decide to install such systems which extract CO_2 or other pollutants artificially from the atmosphere, water or ground. The point is that due to these cleaning technologies, the load on the natural areas is reduced, so we can get closer to balance.

Reducing its territory: today's land usage is pretty rough. I drive quite a lot and look at the land from that perspective. There are wooded strips, bushes along watercourses, flood plains, and unused lands among the lands used intensively by humans. Inside settlements there are also a lot of areas unused or waiting to be built upon. Humankind does not use space in an efficient way, resulting in the habitat and vegetation fragmentation of natural areas. Imagine making use of the agglomeration's territory with 99% efficiency, and giving back the newly-unused areas to Nature! Newer areas are added to the borders of existing natural areas and Nature is allowed to strengthen there. This can be done in such a way as to use the unused areas inside the agglomeration and relocate the land use at the agglomeration's border to inner, unused areas. The result is the gradual shrinking of the agglomeration's territory and the growing of the natural areas around it, which results in the growth of the assimilation capabilities of the natural environment. At the same time, the agglomeration's total and common emissions will be reduced, as the same service level can be reached in a smaller area with less traffic and fewer utilities.

Revitalization: another strategy utilized by the agglomeration can be to help enhancing the assimilation capabilities of the surrounding natural areas with revitalization projects. These can involve water supply, planting, restocking of certain species, etc.

Every agglomeration can combine the above strategies in a way that is optimized to the local conditions—only the global targets are common. Every agglomeration needs different strategies, as the extent of the imbalance between agglomerations and their natural environments is different. The degree of internal emissions and activities, as well as the population density will also be different.

With the optimal combination of these strategies adapted to the local conditions, living in a world which is in balance with Nature can be reached in 35 generations. This system sets Humankind on a developing path gradually. The changes needed to reach balance will restructure and develop the economy, so it will be the driving force of the increasing well-being.

It may be asked why these developmental strategies couldn't be conducted at the level of settlements or nation-states? The answer is simple: the rest of the existing natural areas do not usually correspond with the borders of countries and cities (of course, there can be borders where there is correspondence). As **the cores of Nature's revitalization, these must be kept and be joined by newer and newer areas.**

In other words, the borders of agglomerations must be set according to the existing natural areas. These are the most important factors on which the system is built. These natural cores of areas are the ones from which Nature can recover if we start the work immediately.

5.5. The power of bottom-up initiatives, or: how is this system established?

Agglomerations could be established by setting up a global agglomeration directive which would be accepted by the whole world, and we can simply build from this system. It would be wonderful, but it is not realistic. Politicians around the world would argue about the content of the agglomeration directive for decades while Humankind would reach the edge of collapse.

The system is viable and it will definitely be built up globally. You may rightly ask: but how? In whose interest will it be? Who will start it? As today's world is based on competition, who will take the competitive edge?

The answer is simple but exciting. As you read this book, you will realize that this is the only solution to save the remaining natural values in your environment from final destruction. But it is still not enough to do a lot for it, is it? At the same time, you will also realize that this is the only way for the region where you live to take a competitive edge against other agglomerations in the mid-term. As a matter of fact, migration has already started in the world. **There are more and more shiftless areas. Those regions where there is an unspoiled natural environment and people's happiness level is high will win more recognition in the future.** The agglomeration and revitalization programs provide the framework for it, so the transition has to begin today in order that your region would not float to destruction but be able to keep its natural, cultural and economic values as well as the continuity of its development. What is needed? The collaboration of some agglomerations which are surrounded by remnants of natural areas. This can be realized as a bottom-up initiative as it is not a large-scale step. This collaboration will be the first model agglomeration. They will be the winners as they first apply this system which allows it to save itself. If the first agglomeration will create a buzz, more and more people will think alike. From that point on, the system will spread and develop like wildfire. The more regions that join the system, the more national political power will stand behind it. The international agreements will be developed by confronting good practices until the system becomes global.

The good thing is that the launch of each agglomeration is both a local and world-saving action at the same time. We can therefore develop into an increasing and strengthening climate protection social network, while the rudiments of climate protection can be experienced from the beginning.

5.6. Land use ratio and the modification of agglomeration boundary lines

The land use ratio must be measured in every agglomeration on a yearly basis. This ratio is the proportion of the actively used areas to the whole area of the agglomeration. The target is to have this ratio above 99% in the long term, as I have already mentioned. This ratio is below 50% in an average European area today. Pursuing the objective, areas can be given back to Nature gradually, as a result, **the boundaries of the agglomeration will shrink and the benefit is that we give space to Nature.**

I am frequently asked whether Humankind will have a lack of space due to the areas to be given back to Nature. Area regulation with the land use ratio helps to avoid this situation, since we give back only those areas to Nature which can be freed with the progressive increase of the land use ratio. If the imbalance between an agglomeration and the natural areas surrounding it is still too high, this has to be solved with the other balance strategies, so there will be enough space for humans in this system.

The agglomeration strives to minimize its territory. But what happens if a valuable raw material deposit is discovered under the natural area? In order to exploit it, the border of the agglomeration can certainly be stretched in that direction, but the agglomeration has to give back an area to Nature with the same size and natural values. It is obviously not possible to construct under the natural area from the agglomeration and to get around the system that way. Several past examples can be seen of similar attempts to circumvent regulations. When creating the detailed rules, we have to consider these tricks!

5.7. Traveling and trade among agglomerations

If you travel between Budapest and New York by air, you cause so much CO_2 emission that the Earth is able to bind your other emissions during your whole lifetime. This is to say that with only one such trip, you use up the possible emissions caused by you in your lifetime. Tourism between continents therefore plays an important role in overloading the Earth.

A further problem is that due to global commerce we frequently order products online whose point of shipping—which could be in another part of the world—we do not know. The load on the environment coming from commerce is enormous. The big trick of global commerce is that goods produced in poor countries with cheap labor are sold in the rich parts of the world. As a result, the proportion of the purchased goods produced locally is lower and lower. Due to this and other factors, CO_2 emissions in the transportation sector will grow by 50-250% in 30 years if we do not change this situation somehow (Paul Hawken 2020).

But how? Let's see the solution provided by the agglomeration system.

Obviously, the free shipment of goods and movement of people must be provided between agglomerations. The shipping from realistic distances must be mitigated by taxes. Every form of traveling and shipping must be taxed. The further we travel or transport goods in a more environmentally destructive way, the higher the tax is. For instance, I can travel by plane or by boat to another part of the world. The tax will obviously be higher in the case of flying. At the same time, I can decide whether to go on holiday with my family to a near or far agglomeration. The further we travel, the more tax we pay. This solution does not limit our freedom; it just expects citizens to take the environmental consequences of their deeds.

Due to this system, **local commerce, production and tourism will be boosted again.** The impact on the environment from redundant freight flow stops. People will incorporate environmental awareness thinking into their traveling habits. Companies will be interested in green shipping facilities which will galvanize the development of electricity- and hydrogen-based transportation. If a company uses green transportation for their goods, they won't have to pay as much tax. Companies will also be interested in choosing local or nearby suppliers, so it is a win-win situation! It is true, however, that maximizing profits will not be as inordinate or easy as today in the global market, but we will live on a more livable planet in return. If the introduction and increasing of taxes is progressive and forecasted, the sectors interested in commerce, shipping or tourism can prepare for the transition without any significant economic downturn.

At that point, the counterargument may arise that the above narrows down our freedom and limits our possibilities. This is certainly not true; we can keep traveling anywhere and buy products from distant places. The situation will change in a way that the value of local buying will increase, and using products and services from further away will be less frequent. From this aspect, we will take into consideration more how far to travel. In my opinion, I will not be less happy if I drink Californian wine in Hungary only on special occasions and at other times I drink local (and quite fine, by the way) wine.

5.8. Global Happiness Index (GHI)

In the chapters on the Program of Agglomeration I have talked mainly about the balance of Humankind and Nature. One of the main pillars of the Program is really that the emission-assimilation balance of agglomerations and their surrounding natural areas must be found. However, if you have made it this far in this book, the following statement will be obvious to you: **every form of environment pollution that affects the state of balance and every type of environmental destruction comes from the problems of the human spirit.**

This brings the obvious consequence that without boosting Humankind's happiness, balance with Nature cannot be imagined. That is why the other main pillar of agglomerations is to maximize the Global Happiness Index (GHI), the technical details for which are going to be introduced later in the Program of Happiness. What is important here is to understand the principle itself. The aim of the agglomerations is that people living there should be as happy as possible since if people's average happiness is increasing, social and natural tensions decrease.

Now that we can see and understand this clearly, we can talk about GHI. GHI can be scientifically measured today. We know exactly which factors matter in a person's happiness. These factors can be calculated precisely within a social system, for example in a city, country or agglomeration. At the same time, due to Bhutan's initiative, the UN introduced the calculation of GHI for every country in the world. So, to fulfill this initiative the UN calculates the GHI of every country every year, and you can check the results on the internet. However, the GHI of the UN is totally different from the one suggested by Bhutan. The UN, of course, takes economic advantages into consideration in a very high percentage while the spiritual questions are only considered at a minimal level. In contrast, in Bhutan's suggestion spiritualism gets the focus. So, according to Bhutan's leaders, the greatest global happiness is in Bhutan, while according to the UN Bhutan is only in the mid-range. According to the UN ranking, the Scandinavian countries are the world leaders.

I am not much of a traveler, since I am aware of the immeasurable ecological impact of traveling and at the same time I see clearly today that inner peace and happiness do not depend on chasing new experiences. Traveling is also a way of gathering new experiences, and a lot people raise this habit almost to the level of an addiction. In spite of this, I would really like to get to Bhutan once with the aim to live there for a while. This is not a realistic desire as plenty of things connect me to Hungary. Obviously, this is not a coincidence as I was born here, I have work here, and I have tasks here. But why am I longing for Bhutan's atmosphere?

Bhutan's political leaders are focused on the maximum happiness of the country's population. The main objective is to reach the maximum happiness and to maintain it. The population is mostly Buddhist. What we need to know about Buddhists is that their religion respects Nature to the greatest extent. This is also the reason why it is the closest to my heart, however I do not belong to any religious denominations. Buddhists believe in reincarnation and consider every living creature to be their deceased parents or grandparents, as they cannot know whether or not it is so. In other words, they do not kill or hurt any living creature unless it is indispensable to sustain their lives. They believe that their parents and grandparents would sacrifice their lives happily for their children but as it is an incredibly big gift, it can only be used as a last resort. This mindfulness steeps the society so there is no redundant environment destruction, and apparently people live in more honest and purer relationships with each other. Necessarily, its consequence is simplicity. Society lives in peace and harmony but in a simple way. Financial well-being is a secondary issue. I can illustrate this with an example. A vote was conducted many years ago: the country could save so much money if it installed electricity in every village. Leaflets were handed out regarding the popular vote to inform people how many advantages there were to supplying electricity to every house. At the same time, it was also disclosed that they have money only for installing aerial cables. Aerial cables are however dangerous to birds. Although not many birds die from cables, it is more than none. The country's population voted decisively against the realization of the electricity supply. Respect for Nature is not a joke there!

Because of this, the ecological systems in Bhutan are almost untouched, as people really live in harmony with Nature. This country is a perfect example of the principles introduced in the Revitalization Program and for several other principles in the other Programs. This means a real and almost ready agglomeration already exists on Earth! It is true, however, that there is no democracy in the country, so there are still things to be done there to create a happier future.

Bhutan made a promise at the Copenhagen climate summit in 2009 to remain carbon neutral. At the 2015 Paris climate summit they showed that they had kept their promise (the only country to do so) and promised to keep it in the future. What is more exciting is that they exceeded their promise, as they are the only country with negative carbon balance! They do not equivocate, but act as they speak. The development index of their society is the Gross Global Happiness Index instead of the Gross Domestic Product (GDP). This practical example shows that the grain of social realization already exists. **Balance with Nature and real happiness go hand in hand!** Neither of them can be imagined without the other while Humankind is the dominant species on Earth. Education and health care are absolutely free in this country; even medicines are free. Health and knowledge are subjective rights. Economic development is important to them, but it cannot ruin their culture or destroy their wonderful natural values. A total of 72% of their country is covered by forests, but on the celebration of their prince's birthday they planted 108,000 trees. Their current main targets are the complete elimination of poaching, mining and hunting on all of their natural forest areas.

How far are we from this in 'civilized' Europe? I'll let you answer that question. It is a valid question whether people are happier in Sweden than in Bhutan. I think it depends on what is important to whom. Someone to whom financial security and democracy are priorities would feel better in Sweden. Someone to whom spiritual development, deep human relationships and a pure natural environment are the priorities would feel better in Bhutan.

You can see from this example that the social basis for calculating an agglomeration's GHI already exists—we just have to reach a consensus on how to calculate it. My opinion is that we also need a GHI which takes local specificities into consideration, like the different social systems and the different cultural habits that result in people's different ideas about happiness. It's therefore important that the calculation of GHI should consider the agglomerations' specifics. The big mistake in the UN's calculation method is that it uses a uniform calculation method, while **the idea of happiness in the eastern world is totally different from that in the western world.** The latter is too material and realistic, while the former is more spiritual. Which one is better can be decided by the agglomerations of course, and they can weight it as they want. It is not important to have a uniform calculation method for the agglomerations because competition and comparability are not the aim. **The aim is to be able to measure on the spot whether we are heading the right direction, and so the agglomeration should be honest with itself. The changes in GHI are a mirror that helps to keep our social development going in the right direction.** That is to say, different agglomerations can start in different ways to reach their targets. It will gradually emerge while learning from each other which strategy is the most efficient for each regional, cultural or religious division. This way, we can avoid extremes. It is crucial that no agglomeration can judge others as to why they do things the way they do, so nobody has an exclusive say in how to develop their GHI. The respect for diversity must be developed among the agglomerations, and the GHI is a mirror to show whether they are heading the right way.

5.9. Territories requiring international protection

There are broad natural areas on Earth whose protection the agglomerations should solve together. For this issue, a global environment protection organization must be set up, as I have already mentioned. We need to understand in this chapter that the assimilation capability of these large areas should be divided in proportion among the agglomerations, and the maintenance of the organization should likewise be solved together and in proportion by the agglomerations.

5.10. The effects of introduction

I feel that a future image is taking shape in your mind about how Humankind and Nature can be in balance with each other in a way that the average happiness level of the human race can continuously increase as well. The realization of the two Programs already introduced has several benefits, which I would like to summarize:
- Humankind's pollutant emissions will be reduced globally.
- Nature's assimilation capabilities will increase globally.
- The conscious cooling of the Earth back to the levels before the Industrial Revolution will start
- The new agglomeration system will start a settlement, industrial, commercial and agricultural restructuring which is much more efficient than the current one. This will be the driving force of economic and social development.

- The basis of efficient population control, which is based on truly democratic principles, will be created as the balanced population number can be defined by agglomerations. There is a system built on this which I am going to explain in the Program of Population.
- The restructuring of transport, utilities and settlement structures will start new investments because the discussed interconnection requirements of the agglomeration system totally differ from the current infrastructure system.
- The modernization of agglomerations regarding energy efficiency and energy management will stimulate the economy. Besides, we will live in a more comfortable, modern and eco-efficient environment.
- The system of self-sufficient agglomerations provides a basis for a new global social system which does not build on the competition among nations against each other. I am going to give details about this in the Program of Population.
- The new agglomeration system ensures the revitalization of Nature and the elimination of vegetation and habitat fragmentation, helping the strengthening of biodiversity. It will further boost Nature's assimilation capabilities.
- Nature will recover, which also helps recover the balance between Humankind and Nature and provides the conditions for coexistence.
- The global energy policy will get a proper base and the Global Grid can be established.
- The global interdependence developing in the field of energy policy makes people realize that interdependence is much more efficient than the world of selfish competition in other fields as well.
- The Agglomeration Program will provide a basis for spiritual development and, as we are going to see later, spiritual development will support the creation and strengthening of agglomerations.

- Humankind's GHI will grow, that is to say, the average happiness level of people will grow while unrest will fall.

The following four Programs (Program of Population, Program of Happiness, Program of Society, Program of Economy) will help this transition. The Six Programs, enhancing and supporting each other, will reverse Humankind's development from the current downward spiral into an upward spiral. To reach it, the progressive and parallel phasing in of the Programs is needed.

5.11. 11 Frequently asked questions, counterarguments

How big is an agglomeration, and what would life be like there?
An agglomeration can be any size, as its boundaries will be defined by the existing natural areas. A group of several settlements or a bigger independent settlement can also create a single agglomeration. An agglomeration can be as big as a country or it can be a territory of some villages. Life will not change significantly in the agglomerations compared to life today. People will be happier and live in settlements which develop more dynamically. At the same time, people can still decide between living in a village or a city atmosphere, or what kind of job, living or educational opportunities they choose. Every activity is done within the agglomeration, from agriculture to education. We cannot even take excursions into the natural areas, so the conditions of relaxing and recreation are created within the agglomeration by maintaining parks and fishing lakes.

The significant change will be that we can travel among the agglomerations only on certain high-performance traffic routes, and there will be no network of connecting roads. It is important to point out that roads will still exist within the agglomerations, but not between the agglomerations in order to reduce habitat and vegetation fragmentation.

Will the protection of national cultures and genomes be damaged?

There is nothing to worry about because the boundaries of agglomerations are created by natural areas which are usually big rivers, lakes, mountains, deserts, seas etc. If we observe the territories of human population as genomes and human culture, we will soon realize that they hardly ever depend on borders. Genomes or cultural systems were organized according to natural borders in the past. Along most borders, many people with the same culture and nationality live on both sides of the border. Political forces and power were rarely in parallel with the territorial borders of genomes or local cultural habits. To preserve genomes and local cultures, agglomerations are clearly more suitable than national borders.

Will the Global Grid destroy national energy policies?

A national energy policy tries to maximize national energy safety and gain an economic competitive edge among the other countries by providing the cheapest energy. A national energy policy is also interested in maximizing energy exports since these boost the national economy. The social-economic result of this way of thinking is that every nation maximizes its energy investments, so the installed over-capacities are dramatically big globally. But if the Global Grid is realized, electricity will cost about the same everywhere in the world; furthermore, in the long term it could be free. That is to say, energy will neither be a barrier to development nor a tool of competition. That is why overbuilding a country's energy supply is useless. The principle of interdependence enables the establishment of the most efficient systems, which can provide the highest level of service with the lowest amortization and for the cheapest price. As a result, Humankind can use its capacities to develop other fields. Globally cheap or free energy will not be the barrier to development; the Global Grid instead becomes the driving force of development. We need to admit now that energy policy cannot be handled at the level of national competition anymore—an out-of-date, obsolete system which is the obstacle to our development and also the obstacle to the global fight against climate change. Every country fulfills its commitments, but in fact every country tries to minimize its real acts because they do not want to make their energy systems more expensive than those of other countries—once again bringing us to the fact that national competition restrains the chances of efficient intervention.

Does the growth of natural areas cause any tensions regarding private properties?

As we want to give back lands to Nature which are declared to be natural areas, ownership will be inevitably violated. Humans cannot possess the whole Earth in the future. The obvious result is that companies and individuals possessing these lands may be prejudiced. It is the government's task to compensate the owners for losing properties that are on natural areas with same size or bigger area within the agglomeration, or with money. If the owner receives an area within the agglomeration instead of an outer natural area, it will be a good compensation. The introduction is of course gradual. First, the existing natural areas become the core of the future natural areas, which are usually state or municipal properties. During the gradual expansion of these areas it may be necessary to expropriate lands for the sake of natural areas. Fair compensation becomes important in these cases. How will it be financed? I am going to answer this in the Program of Economy.

Are national identity and respect for my culture outdated issues?

By no means! Insisting on retaining national culture is a help and support for the transition to the Program of Agglomeration. An agglomeration is a more suitable system for protecting local culture and language. The agglomeration system therefore brings with it the further development of national thinking.

Will there be enough economic power for the transition to the agglomeration system? How?

When creating the agglomeration boundaries, it is not just the existing natural areas to be taken into consideration but the existing infrastructure. The already-functioning highways, sewage farms, and drinking water networks can be used. The great thing about this system is that we do not have to rebuild the whole world, but are able to change gradually with properly managed technical planning.

If we do not hunt on natural areas or intervene in other ways, there will be too many of certain animals and it will cause an imbalance! How can we control the increased number of animals if they break into the agglomeration?

If we leave a natural area alone, the top predators will move back, so hunting will be needed only in the temporary transition period. Where top predators have died out, there they will have to be re-established. The protection of agglomerations can easily be solved by electric fences, drone surveillance or with several other types of technical equipment.

Why can't we take into consideration the assimilation capabilities within the agglomeration when calculating balance?

The engineering world calls the ignoring of assimilation capabilities 'neglecting for security'. As there is much uncertainty in calculation methods and people are likely to exaggerate their social performance (at least until reaching a certain spiritual development), a safety margin is needed.

Doesn't separating the agglomeration and the natural areas make the mistake of isolating Humankind from Nature forever?

The aim of social restructuring regarding agglomerations and natural areas is not to destroy Nature but to give space for it to recover. This solution will teach us to respect Nature! When we have reached a balance between Nature and humans on Earth, we can gradually loosen the sharp borders between agglomerations and natural areas. But this would take place in the distant future—beyond the scope of this book.

CHAPTER 6

The third program of change: the Program of Population

6.1. Oops: we're overpopulated

I am convinced that the world is overpopulated, but this is not a new idea. Before I deal with the solution, I would like to show you how serious the problem is. There are several things to support the statement that we are overpopulated: the reduction of biodiversity, the dramatic loss of natural living areas, the significant reduction of ocean fish stocks (in many species only 3% have survived), the incredibly fast destruction of rain forests, and the accelerating climate change. The situation is clear: human beings have become the most powerful predator, and their selfishness and covetousness have reached unprecedented levels due to their modern-age spiritual distortion. This is enhanced by the overabundance of population and the extreme desire and ability to rule Nature that goes along with technical development. Now add the issue of population to this complex problem.

If we imagine a natural community as it developed gradually during evolution to strike a balance, we will see there is one top predator in hundreds of hectares of natural areas. It doesn't matter whether we observe the relative number of lions on the savanna, or of bears in taiga forests, or of sharks in the ocean in a given area: we will get similar results. If we accept the statement that Humankind is a top predator with much higher needs than any other top predator, space per person should be bigger than for lions or bears without upsetting the natural balance. To make it simpler: the average density of people should be less than the natural density of lions. As we have already exterminated a large portion of the predators on Earth, we took their place (for example, bears no longer exist in Hungary and wolves rarely appear, although they were once the top predator), so we can add their previous numbers to the maximum population number that could be reached.

If we continue this way of thinking and we project it onto the whole Earth, only about 800 million people can live on Earth according to my estimation. This is one-tenth of today's human population. Of course, this is only a rough estimate: it could be determined more precisely scientifically. If we take into consideration the efficiency gains coming from the development of technology, this number could not be much higher because human comfort needs are much higher than those of any other top predator. If we start from Humankind's carbon footprint, it turns out that if everyone lived at the same standard as people in Germany, we would need 4 to 5 Earths. If we extrapolate back, Earth would be able to support a maximum of 2 billion People in such a way that their needs are expected to satisfy the western model, considering basic human dignity.

To sum up, the Earth's capability to support humans is somewhere between 800 million and 2 billion people at the current technical and social level and at our actual level of consciousness. It is crucial to highlight here that I'm not stating that only that amount of people have the right to live on Earth; every individual has the same right to live here. Indeed, we can say that Earth would be able to support 20 billion people without any problems if we didn't live the way we do now.

The population that can therefore be supported will change dynamically in the future because the growth of the technical-economic level will increase this number. Humankind's spiritual development has the same impact. In contrast, the destruction of environment decreases this number by decreasing the carrying capacity. **There is still hope in the future that technological development can proceed to find a balance with its environment without any further damage to the natural environment.** If so, the carrying capacity of Earth will improve and it will be able to support even more people at a higher standard of living without sacrificing balance. This book is designed to lead us in that direction. On the other hand, if Humankind carries on following its current developmental tendencies, the carrying capacity of the natural environment will keep on decreasing at a dramatic pace, which could not be compensated by technological and economic development. The consequence is the decrease of Earth's carrying capacity, and later the breakdown of our development and sudden, torturing population loss. Our extinction is a real possibility. But before thinking that this possibility is far in the future, unfortunately I have bad news for you: the big fall will begin in about 2040 if we do not change. And it could be even sooner.

Simply put, if we do not want to get there, we are in the final moments to make a change. It's our decision which of our present options we will choose, but the situation is urgent.

6.2. How can we solve overpopulation democratically?

Before I get started, let's set the target for the Program of Population. This is the **mid-term target for reaching a balanced population number.** A balanced population number is the global population that can be supported by the natural environment at a given social and economic level without losing balance with Nature.

The current population number is many times more than the balance number, but the target is to reach it, as this can be the only basis for Humankind's long-term development. A temporary decrease in the global population is also necessary to get to this better future.

The balance population number is a dynamically changing quantity that should be recalculated every year. The calculation is easy: every agglomeration has to calculate its actual characteristic balance population number based on its local emissions and the assimilation capabilities of its local environment. If we sum these quantities globally, we get the global balance population number. If we compare these results every year with the current one, we will have a reference framework and will be able to create short-, mid-, and even long-term strategies.

But to return to the solution, how can we get there democratically, and what kind of strategic assets are available?

I have been thinking about a total democratic solution to overpopulation for about 25 years, and the big picture has slowly come to light. Population reduction has three rules:
   I. Contraception is a subjective right.
   II. Female empowerment must be strengthened and women should have full access to education.
   III. You can have as many children as to whom you can provide the equivalent of at least one adult's continuous, **selfless** attention.

It may seem strange, but this is enough to first slow down the population increase, then after a short peak for the population to start falling. What is more important, well-being will grow as a result, peace will thrive and the average level of happiness will grow. Let's have a look at why.

According to a survey in the USA, 45% of children born there are unexpected (Paul Hawken 2020). Unfortunately, I have not read similar surveys from other countries, but if this number is so high in such a developed country, we can safely say that in Earth's poorer regions this number is higher. To estimate roughly, this figure could be 60-70% globally, which would mean that on average every six or seven newborn babies out of ten were not expected by their parents.

Today it is a scientifically proven fact that if the mother does not connect emotionally to the developing baby with enough love during the gestational age, the baby will be born with a badly wounded spirit (Orvos-Tóth 2018). This is why a lot of us feel there is something wrong deep inside, even though everything is perfect on the surface. We unconsciously carry the shame we felt before birth, specifically, that we are not good enough for our mothers to love us. According to this, 60 to 70% of the population live their lives with a frustration that most of them cannot get rid of their whole lives. Since parents feel ashamed that their children were not planned, they usually keep it a secret. Consequently, most adults carry this really deep spiritual burden without being aware of it, although frustration, inner emptiness and addictions stemming from it become part of their lives.

Today, it is a proven fact that **financing free contraception as a subjective right globally costs only a fraction of what it saves in climate change damage and growing social tensions** (Paul Hawken 2020). Automatic contraception can be easily solved by free pills and free condom machines, or even a quota per person scheme could work as well above the age of 18, sent freely to all homes on a regular basis. To reach efficiency, this scheme must be promoted in public service announcements and raise public awareness in parallel. As a matter of fact, most unplanned children are born because of the lack of contraception. With this rule, the number of unplanned newborns will shrink dramatically on a global level within a few years.

This covers the first rule; let's address the second one. For us in Europe, female empowerment is a common issue and it is generally believed that women have the right to be qualified and educated. This is far from being the case in the bigger part of the world. Women are uneducated in a lot countries and regions; their only task is to do housework and to serve men. Due to this, many of them are unfortunately exploited by men's sexual wishes, resulting in a lot of unplanned children. There are of course loving families in those regions as well, where parents consider every child as a fruit of their love. But we have to admit this is not common. An extensive program was conducted in Ethiopia in which through the education of women it was achieved that women taking part in the program had only 2 children instead of the usual 8 or 10. As a result, this tendency started in the bigger cities in Ethiopia. In Europe, the everyday situation is that a three-child family is considered to be large, while in Africa this is a small family. The emancipation of women has an enormous impact on the population number. If a woman is educated and does not depend on a man, she is able to make right decisions, so all of us have to work on all fronts for female empowerment! In doing so we can do an incredible amount against climate change and to reduce social tensions.

The third rule is the most important and requires the most complex way of thinking at the same time. The basic concept is not mine; it is from *The Celestine Prophecy: An Adventure* (author: James Redfield), which is a standard work of spiritual literature. It is an especially joyful novel and provides deep spiritual lessons, so I happily recommend it to you if you have not read it yet. The author wrote in this book that the only rule for raising children properly is to have as many children as to whom we can continuously provide the **permanent, selfless attention of an adult** (only when the children are awake, of course). We have to provide this until the age of 12 to 14, as at the beginning of adolescence children detach from their parents, and the 'attention obligation' starts to reduce from that point.

It also belongs to the interpretation of the rule that this 100% adult attention does not necessarily come only from the parents; grandparents, great-grandparents, relatives, and friends can be involved. The point is that the children get loving and selfless attention all the time. What will be the result?

The result will be that by the time of their adolescence, a deep primordial trust will have developed in the child and they will be aware of their own specialness and uniqueness. Children will have a deep self-awareness and accept themselves as they are. Just imagine: if you had reached the age of 12 or 14 with these conditions, what could have become of you by now? Unfortunately, only 1 to 2% of children today and most adults do not attain this condition of deep self-acceptance until the end of their lives. Imagine what it would be like to live in a society where peaceful, self-accepting, non-addicted people walk the streets, or you are able to work with such people at your workplace! I know we are far from having this today, but Humankind is able to move in that direction gradually. To reach it, we just have to change the basic pattern of our thoughts.

Parents today do not have time for their children. An average European parent has 10 minutes of totally selfless attention for their child on a daily basis, but a lot of other parents have even less. Most parents are thinking about something else while they are supposedly paying attention to their children. Children see with their souls, so they feel when a parent pays attention only superficially. From this experience it builds into their unconsciousness that they are worthless and do not deserve attention. This is one of the main reasons why there are so many spiritually damaged people and so many frustrated children with behavioral problems at schools. Surely it is starting to become clear why this rule is so important, isn't it?

This rule is double-edged, and that is why it is so efficient. Most prospective parents want the best for their children, so if this rule is embedded in the common consciousness (promoted and taught), every prospective parent will realize what they must do if they want happy children. Because of the need to provide attention, every couple will think about the number of children they want to have. This way, families will have fewer children but they will raise those children to a much higher developmental level.

And here comes the other perhaps even bigger impact of this rule: **the first generation will appear on Earth in the near future where the proportion of self-accepting people is greater than those who do not accept themselves.** Imagine how big a move this will be regarding the average level of consciousness! This generation will have a totally different approach to consumption, economics, peace, respect for Life, or any other important question than us, as our destruction of Nature and social tensions essentially come from our spiritual wounds. For this new generation, the behavior of people today will be absolutely incomprehensible.

The aim is to reach the balance population number with the help of the three rules of population control, which means that the emissions and impact caused by Humankind will be in balance with Nature's carrying and assimilation capabilities. These rules not only help in reaching the population number, but ground the success of the Program of Happiness and the Program of Population and Economy, which will come soon.

6.3. Society's phobia of aging

It follows from current economic principles that we keep ourselves in a constant force of economic growth. If the economic growth is 'only' 2%, we are talking about recession, although this level means that the quantity of goods produced doubles in 35 years. In other words, in two generations the average standard of living grows about two times more. Due to the chasing of economic growth, population reduction is the economists' nightmare. When a population is shrinking, a growing percentage of the population is retired. The population shows an aging tendency and as a result, it is difficult to maintain economic growth as the proportion of the workforce actively taking part in producing GDP growth becomes disadvantageous.

As a result, governments don't want to hear about population decline. (Exceptions are the countries where the population growth is so high that the economic development is not able to follow, like in India nowadays.) Where population reduction is experienced, everything is done to increase birth rates. The aging EU is like this, but we have to accept that at a certain level of social spiritual development it is a natural process. The governments of EU countries have tried in vain to encourage people to have more babies; there has hardly been any success in the last few decades. There are several reasons why people in certain regions want to have fewer children. One of them is that financial well-being reduces the survival instinct, and because of this effect people have fewer children. The reversed interpretation is that if survival is no longer a driving force, Humankind will be able to live at a higher spiritual level, and at that higher level other preferences will appear. People give more attention to fewer children and use more resources for fewer children in order to provide better chances to start their futures. So those principles that I introduced as basic concepts of population control a few pages ago seem to have spontaneously begun within society. There is no problem with the reducing populations of nations; it is a natural, normal process! If we hinder this process, we do so against our own mid-term development.

Population reduction can be easily solved in western countries by helping the settlement of immigrants. There is no reason to be worried about lack of workforce or economic recession in the mid-term—not even regarding the Population Program!

The other big obstacle is nationalist thinking. Neither national policies nor the strengthening of national identity allows for population decline. Why? It's simple: when the population declines, politicians start to frighten citizens about the weakening or the end of their nation. However, if a 0.5% population decline starts in a country, it will not mean the disappearance of that nation, even in hundreds of years. What happens if a population falls by half? In the case of a nation with a current population of 10 million and 5 million in the future, the reduced population will still be able to maintain national culture, language, and anything else that's important from the national aspect. Of course, people can be afraid that with a smaller population it will be much easier for evil and bad foreign nations to steal theirs, but we have to admit to ourselves that this is an outmoded way of thinking. World peace is at least not impossible, and it will definitely come if Humankind undertakes the right way of its development.

The other counterargument for frightening people with population decline is that there are huge differences in the performance of individual countries and regions, as this has been already supported by a lot of sociological, economic and social analysis. Solving the economic problems occurring due to population decline can be achieved by increasing the average level of knowledge and enhancing workplace efficiency (mechanization, automation, technical developments, etc.). With one word, the improvement of the average standard of living can be realized—even in an aging society without a flexible immigration policy.

6.4. Religions and contraception

The roles and responsibilities of religions are huge. Most religions do not want to hear about contraception! But unfortunately, religious dogma is usually the obstacle of development. The religious slogan 'become many and fill the earth' may have been proper 2000 years ago, but by now we have accomplished it—in fact, we have over-accomplished it. It is therefore time for some religions to renew themselves and help Humankind to keep its direction to solve climate change. I have to point out that I deeply agree with the basic values delivered by the big world religions; I am not against being religious, but rather I'm against exaggerated dogmatic extremes which often block development.

An ethical reaction is frequently heard from religious people that every conception is made by God, so we cannot decide the number of babies to have. In my opinion, this is the perfect manifestation of religious dogmatism socially. Certainly, we cannot decide who shall be born, so we agree on that! But we do not agree on the notion that a mother cannot decide with contraception when she is ready spiritually to have a baby.

6.5. Families and childcare hardships

Families have real difficulties if they want to keep to the principle to always have an adult with full attention on their child. The current education system is not capable of this, and neither is the social system (details in the Program of Society.) But if we realize how important it is, we can head in this direction gradually. For example, more and more mothers might choose to stay at home on their own accord for the sake of her children's better spiritual development. The great difference from the older family model is that more and more mothers may choose to do this freely in order to live their motherhood in a way they prefer or is beneficial to their children, especially if the father or grandparents are unavailable or unable to perform childcare to an acceptable level. The process has already begun in certain levels of society but it is not a mass phenomenon yet. The result of the European emancipation is that women are able to earn their living so they do not depend on men financially. That is why men generally cannot rule over women as much as in the past or in other societies.

6.6. The main effects of the Program

The results following the basic rules of population control:
- First stabilizing, then gradually reducing population to the balance population number.
- More balanced people, fewer addictions, less needless consumption.
- Growing of resources per capita, reduction of unhealthy competition between people due to less apparent limitation of available goods.
- Decrease of environment pollution.
- The average standard of living grows.
- Less social tension.
- The revitalization of Nature is boosted.
- Female empowerment rises in the world.

It should be clear that this Program is related to all the other Programs and sets Humankind's development globally in the right direction.

The point of the solution here is again a phased-in model, but the most important thing is to dare at a political level to set population decline as a long-term target spanning generations. First, we need to reach only in the direction of stagnation. Future generations' spiritual education is needed to reach population decline, especially since people have to accept the rules and principles of child care already introduced. To succeed, multigenerational education is needed to offset the weird human tendency to maintain old habits to their limits—even when they turn out to be wrong or defective. The reasons are to be found in how the ego works, because if an idea (e.g., religious dogma) becomes part of our ego, the ego will do everything to keep it so. You can read more about the ego in appendix 1 of this book. I also highly recommend Eckhart Tolle: *A New Earth* regarding the topic.

At the same time, the fight for female empowerment has a long history. However, if we make people understand that it is also one of the most efficient tools in the fight against climate change, we will be able to achieve bigger successes with it.

Introducing contraception as a subjective right globally (or by countries one-by-one) will alone stop the global population growth in a few years!

CHAPTER 7

The fourth Program of change: the Program of Happiness

As you already know, only a balanced and happy Humankind can live in perfect balance with Nature. The aim of this program is obvious: the greatest possible boost to human happiness, as this is the most efficient climate protection as well. But I would like to highlight again that by 'happiness' I mean the condition belonging to the Life-supportive levels of consciousness, and not the immoderate chasing after of material goods and services as already discussed in Chapter 2.

I have explained the interpretation and calculation of the GHI in previous chapters. The main target of this program is to raise human happiness to the highest level. We can express this in numbers with the GHI value.

7.1. Selflessness, community life and the GHI

It is also obvious that immoderate selfishness at the individual, community, national and regional levels is the destroyer of the world. In contrast, boosting selflessness raises Humankind's average level of happiness at every level.

Selfishness breaks society into lonely individuals. It is not by coincidence that the basic component of society falls apart, namely the family. The individual in themselves is much more vulnerable than in a community, even though the ego suggests otherwise (see Appendix 1 for details), so any form of community life combats this.

A social demand has started within this decay to motivate social movements for social purposes; this can be seen on a daily basis among the aims of EU competitions. It is difficult to get people to move because self-interest is more important to them than community Life, but this is only the present! The aim is to mitigate selfishness and strengthen the feeling of togetherness, selflessness and community approach. If we move in this direction socially, it will have beneficial consequences in a few decades. With the decline of selfishness, the GHI will grow in inverse proportion.

It is crucial to strongly deal with the questions of selflessness and social togetherness by developing the approach.

There are already initiatives for such processes in society. A lot of examples are known where civil organizations have achieved major success even against large global companies.

7.2. Media Ethic Codex

The state of our spirit strongly depends on whether we engage in activities that ruin or build it. The opportunity to choose is given at every act, but we usually decide unconsciously without being aware of the spiritual consequences. From this aspect, media has a large impact on raising or lowering Humankind's GHI. Unfortunately, the mainstream media is ruining GHI dramatically. In order to be able to speak about a solution, we should first see what the reason is exactly.

People today switch off by watching movies and series. If we consider what is worth watching from our spirit's perspective, it's far from certain that we would make the same choices that we would for entertainment purposes.

We've been watching lots of American-style action movies for decades. The plot is always the same. There's a hero who is angry with some invincible evil force, but the hero destroys that evil and saves Humankind. There is always of course a love or erotic element in the movie, and the hero usually loses some relatives or friends; the hero pays a high price for victory. The never-failing recipe is therefore a positive hero, plenty of action, fights, a lot of murders, torture, struggle, love-eroticism, exciting visions, and finally the victory of the good side.

Let's have a look at this type of movie from the perspective of the levels of consciousness. We get hooked by the whole movie or by some parts of it because we can really enjoy a movie if we can identify with the story. Our spirits will work at the level delivered by the movie. In about 60% of the time during a typical action movie, the good and the bad sides fight with each other. Tough wars, murders and fights are going on. The level of consciousness of these scenes is Opposition (Anger), whose energetic value is 150. The movie depicts a good ideal inside and outside. The hero's love is usually the most beautiful woman. The emotion between the hero and the loved woman is the perfect love itself. These ideas as shown by the movie push our spirits to the level of Desire without noticing it during 20% of the movie; its energetic level is 125. In order to deeply identify with the good in the movie, the director has to show the evil as being infinitely bad and scary. In these scenes, our level of consciousness reduces to the level of Fear, Shame and Guilt, whose energetic value is at maximum 100. This is usually 15% of the movie's duration. When good wins and we experience catharsis, it supports the feeling that we are on the right side and raises us at the end of the movie to the level of Pride—whose energetic level is 175—for about 5% of the movie's length. If we take the weighted average of these levels and time periods, we get an average level of 140, which is a strongly Life-destructive value. To solve climate change, we should be reaching a level above 200, but these movies drag us in the opposite direction on a daily basis. **Big media has such a big responsibility, but the media today takes no responsibility for the consequences of its actions!**

See what this means from our spirit's perspective? It means that we went to the cinema and paid for falling to the fairly Life-destructive level between Anger and Desire, and ruining our spirit for an hour and a half. I watched these movies for decades and didn't understand why I felt it wasn't right, because it should have been fun. Why did I feel that these American-type action movies weren't quite right? I couldn't explain it; I just knew it was not okay. Now the answer is quite obvious, isn't it?

These movies have another element, which is called the negation of evil (J. Bradshaw 2015). This is the ethical strand of the movies. In these movies there are no differences between good and bad in terms of their actions. The good side also destroys and kills, but they are beyond every human and ethical rule because they are on the good side in the movie, and they can therefore do everything in the fight for good against evil.

Unfortunately, we identify ourselves with the evil in the movies too, all the while negating it. This is the most insidious spiritual trap; we do not even realize that we are on the bad side. Consequently, in these movies there is no good or bad side, as both sides are effectively evil but one is shown as being good. If I kill for the sake of good, I am still a murderer. Most of the heroes are common mass murderers. Are they 'good' just because they stand on the good side?

If you add the logic of levels of consciousness to the negation of evil, it becomes crystal clear why it's so bad to watch these kinds of movies, and understandable why this activity ruins the spirit all the time.

At this point, a lot of people say: 'Come on—I seen lots of similar movies and they don't have an impact on my spirit". Unfortunately, this is the ego talking; we always try to justify our past decisions. **It is easier to believe that we and most others are doing the right thing** than to believe the opposite. It takes courage to challenge the rightness of our decisions.

But there's no need to stop our analysis at simple American-type action movies. Let's see how much harm we cause to our spirits by regularly watching today's average media products:

- Advertisements: level of Desire, energetic level 125.
- Horror movies: levels from Fear to Shame, energetic level 50.
- Sex, erotic movies: level of Desire, energetic level 125.
- Thrillers: level of Fear, energetic level 100.
- Hard porn movies: levels of Shame, Guilt, Desire, energetic level about 50.
- Intensely romantic movies: level of Desire, energetic level 125.
- Action movies: levels from Pride to Fear, energetic level 140.
- News: wide range of levels from Guilt to Anger, average energetic level 90.

- Nature and educational movies: level of neutrality, energetic level 250.

If you consider how many ads and spirit-destructive programs are broadcast to you by TV channels, you will understand how after watching TV for one or two hours you get up feeling lazy and drained of energy. Your spirit was literally depleted by TV.

Once I watched a popular Hungarian TV channel to see how many programs or parts on a given day raised us to a Life-supportive level of consciousness. At most, I found only one hour of programming like that out of the 24 hours. This gave me my reason to give up watching TV. Now I hardly ever switch it on. Under the guise of entertainment, our spirits are ruined continuously through TV channels. Do we need it? Of course not! Spiritual development is challenging enough without media like TV ruining the effort.

The world of the internet is even tougher because we can get hooked by the most exciting temptations with only a few clicks. These usually drag us down and are not uplifting things. It is always too easy to fall back and too difficult to move on in spiritual development. In this case it is not the easiest way that is the right one.

I usually get another reaction to my thoughts about TV: 'If that was true, people wouldn't enjoy these movies and programs so much'. This is also the reaction of the ego, which wants to justify our habits. It is an instinctive and natural reaction, but it is not necessarily correct. Everything depends on the basic level of consciousness. If someone lives at the level of Guilt, they will experience a positive catharsis even in the case of a typical American-type action movie, as their energetic level of 50 will be raised to 140 temporarily. They will come out of the cinema full of enthusiasm. That is great. In this way, watching action movies will be a significant help in their spiritual development. Go ahead! Watch a lot of them! If someone lives at the level of Desire, they cannot recognize that porn movies and ads drag down their spirits because they are on the same level, so it does not lower their spirit but keeps on the same level. But from the levels above 300 the differences can be clearly felt. So even though you don't feel it the problem is there, and the fact that you don't feel it underscores the need for spiritual development. You can't really be happy living on a Life-destructive level. It is true however, that we have to first look inside ourselves honestly to see how happy we are.

Today a lot of people live on Life-destructive levels much more than on Life-supportive levels. Media serve the crowd, as the media are only interested in viewership figures. Unfortunately, they don't care about hindering people's spiritual development on a daily basis. Today, the mass media is one of the greatest barriers to Humankind's spiritual development. Of course, more Life-supportive programs—and also entire channels—are emerging. This clearly shows that change has already started. These channels have lower viewing figures though, as they appeal to a smaller audience (so far). But if we realize and are aware of the problem, we will generate change in our lives, and media products will change as well. The media industry serves us; the way it develops and changes depends on our decisions.

At the same time, why is it easier to click on a spirit-destructive series or action movie instead of reading a spirit-lifting blog? The answer is easy: developing the spirit is the same as training the body. Working hard and keeping ourselves fit is challenging, but lying in front of the TV eating chips is easy. After hard exercise we are full of good sensations. After TV and chips we feel lazy and lacking in energy. We feel like doing nothing. Our spirits have to be trained as well, and this requires hard work and attention. Climbing a mountain is also hard, but the higher you go the better you feel and the wider view you have. The same is true with the levels of consciousness and spiritual development: it's a metaphorical mountain climb. The wonderful feeling a climber gets when reaching a peak is available to you when your spirit rises. It's worth working for it!

This chapter doesn't want to suggest that you should stop watching spirit-destructive things, as this kind of hard-line approach will not succeed in the long term. I admit that l cheat myself from time to time. Sometimes it feels good to be a little bad, or to identify with some bad thing. This book rather wants to urge you to maintain a balance. Try to find more Life-supportive entertainment, and ways to switch off what is spirit-destructive. If the balance is positive, the change in you will start in the right direction. For example, according to the kinesiological measurement of a spiritual master **the energetic value of this book is 703.** This means two things from your perspective: one is that your level of consciousness has risen because you have read this book up to this point. The other is that **this book has a mission to impact Humankind;** many ancient Christian, Buddhist, and Islamic documents have similar energetic values. So if you choose activities with a high level of consciousness, they will lift you up and make you more balanced.

Now you can understand why the media and film industries distort people's spirits and thoughts so deeply. Anything can be done in media that brings profits; ethical, religious and spiritual norms cannot slow it down. The products of the media wash over people's spirits and consciousness completely. Media are only about competing and pretentiousness. The aim is to have the latest and biggest effect on viewers. Meanwhile, these sources push people's level of consciousness downward incredibly strongly. Ads do the same, but with even more insidious psychological techniques. What we think are **our own desires have only been planted in us through media.** If you don't believe me, just think of your desires when you were a child—when your spirit was uncontaminated. You will soon come to realize what brainwashing you have been through since then.

The other very seriously society- and climate-destructive effect of media is the bolstering of individual selfishness. You hear it everywhere that someone can obtain everything they want, selfishness is great, and that an individual's desires are the most important things. **This is a happiness trap that actually takes away the last chance at happiness!** It is nonetheless easy to believe because it is always easier to choose the simpler way.

The solution is to change these trends. Change will be determined by your personal consuming decisions, with which you will boost your happiness and work against climate change and other social problems. At the same time, a **Media Ethical Codex** should be established right away. There have been earlier attempts to do this. It is not obligatory to join this codex, but doing so can provide certain political, social and economic benefits. The different companies and organizations should be involved by motivation; that way the change will be successful. Life-destructive sources will shrink, and Life-supportive ones will expand. The downward spiral of media will turn upward.

7.3. The method of Life-supportive balance

The **promotion of Life-supportive activities and their impact on happiness** must receive an emphasized position in our approach to development and education. Helping and giving are uplifting and healing for the soul, thereby becoming the foundation for the way to happiness. If people get proper information about it, they will become motivated, since reaching individual happiness is the aim of every normal person.

This is what Life-supportive balance is good for. The point is that we can add up all our Life-supportive and Life-destructive activities every day and draw a conclusion about how properly we have lived that day. Obviously, it's necessary to be well informed because we don't usually consider how Life-destructive our activities are. For example, until I started to meditate, I was a big meat eater and I was convinced that it was healthy and appropriate. If it came to my mind that some poor cow, sheep, pig, chicken, rabbit etc. that I was eating had died because of me, I calmed myself down with the idea that they had been bred to be slaughtered. My spirit cannot accept this compromise anymore. I am absolutely not ready to link the killing of animals to my spirit. Of course, I don't want to persuade anyone to quit eating meat; everybody should eat what is good for them. Under certain levels of consciousness there are no such inner needs in people and there is nothing wrong with it on those levels. I just want show that in order to judge Life-supportive and Life-destructive activities objectively, we definitely need more information on the topic.

Therefore, we have to place the information about Life supportiveness into the world of products and services. Details are given about this in the Program of Society. If the Life-supportive or Life-destructive value of products and services can be seen, **the individual will be given the moral opportunity to make the right decision.** Producers and service providers are currently hiding this information. The whole system is sparkling on the surface but rotten inside. **The obligation to inform People will be what starts to clear up the system.**

7.4. Introducing a general mental hygiene system and social prevention, and the practical measure of GHI

People must be supported by free consultancy and psychological help in their search for happiness. Proper forums and opportunities within society must be established for that reason.

I used to deal a lot with psychosomatic illness. I am convinced that 80 to 90% of people's diseases are psychosomatic. Regarding this topic, Joe Dispenza introduces a system based on scientifically proven facts in his book *Becoming Supernatural*, which is one of the best books I've read in my life, so I truly recommend it to you.

In some cases, spiritual problems cause diseases directly; in other cases they cause diseases indirectly. It should go without saying that a happy person has a stronger immune system, so an individual's search for happiness has a direct economic impact as well. One of the biggest problems of modern societies is to maintain its highly expensive health care system. The growth in the average individual happiness level reduces health care costs and social burdens, thus improving economic efficiency (less sick leave etc.). Less medicine and medical equipment will need to be produced, resulting in less global emission.

A legitimate conclusion is that the general practice system must be augmented by a **general mental hygiene system.** A mental health care expert should be employed full-time in every district; their office must be next to the general practitioner's office. Furthermore, there should be a supervisor or coach at every workplace who supports the professional side of people, as we spend the bigger part of our active lives at our workplaces.

The district mental health expert is responsible for the improvement of people's mental health in their district, while the general practitioner is responsible for their physical health. They naturally support each other in their work so that society can move forward. The famous saying 'a sound mind in a sound body' will make sense in this way. This step could be made even more effective with the expansion of the already-working Japanese model. Specifically, the fewer patients general practitioners have in their district, the higher compensation they get. So, the mental hygiene expert and the GP will be interested in doing real healing, and not only the mass treatment of symptoms.

The general mental hygiene system has four main tasks:
- Healing of the soul
- Professional support of an individual's search for happiness
- GHI measuring on a practical level
- Social prevention

We can clearly understand the first two tasks of the mental hygiene system. Now we will focus on the other two.

It is important to establish a system to measure GHI, whose basis is the measurability of the levels of consciousness. The existence of levels of consciousness aids us greatly in measuring who is how happy. By measuring the levels of public consciousness, the level of happiness developing in a given agglomeration can be easily tracked. This represents the establishment of a general mental hygiene system, which can be the foundation of regular measuring. To measure one's level of consciousness takes only 2 to 4 minutes with kinesiological methods. If we visit a mental hygiene expert experienced in kinesiology for 10 minutes a year to measure our level of consciousness, and they then enter it into a database, the average annual level of consciousness will be generated automatically in every agglomeration. This is perfect for measuring GHI, since there is no lie in it. Do you remember the chapter that mentioned that the body is incapable of lying? The good news is that this system is not unrealistic, as there are already experts today. I know one personally; she is currently helping me with my spiritual development.

This value must be the most weighted within the GHI so society cannot cheat itself about how happiness is developing in the agglomeration. That data will soon reveal whether the agglomeration leaders are doing a good job! **From that point on there will be no equivocation in society; we will no longer make reality seem nicer, as is done in today's political situation. It will be a truly honest system,** which will teach us to look in the mirror and be frank with ourselves both individually and collectively.

Let us go back to what was said before regarding individual happiness. If someone goes to the local mental hygiene expert and the measurement proves whether their level of consciousness has increased or decreased compared to the previous year, they will get a reflection of their personal development. **This will provide the motivation for society to take spiritual development seriously and bring it to the fore.** This single step will change the world dramatically, and this single step will reduce global emissions significantly and quickly. **I consider this to be the most efficient tool for climate protection to develop in the fastest way.** But this is not the only advantage of the system: a really serious social benefit is coming next.

It is important to highlight that this yearly measurement does not violate any personal rights! It's equally routine for me to go to my GP once a year for a general check-up. My own measured values are entered into a central averaging database anonymously so there is no chance of misuse. If I was warned by the mental hygiene expert that my level of consciousness had gone down a lot, but I didn't care, that's my decision. No one else has a say in it.

Do you remember the chapter discussing the group dynamics of the level of consciousness? You might have already drawn the conclusion that if there are some people with really low levels of consciousness living in the same city, they will significantly push down the average level of happiness in that city. The reason is not only the fact that their low level decreases the average level numerically in the statistics, as this vanishes from the perspective of math. It derives from the group dynamics of the level of consciousness that someone at a really low level can neutralize the positive impact of 10,000 people at the level of courage. It's no accident that where there are lots of criminals, the average level of happiness is really low. These people cause unhappiness unconsciously even in places where they do not commit crimes. Most criminals are people with really low levels of consciousness, though this is not universal. It can happen that people with a really low level have not committed any serious criminal acts, but they nonetheless push down the level of consciousness in their environment dramatically and reduce their chance in developing their own happiness level.

Here's the issue of social screening: prevention. In today's selfish society where everybody deals only with their own issues, extreme social oscillations are unfortunately more frequent. These can be prevented by the general mental hygiene system. A lot of social tensions, criminal acts and the load on the environment coming from these things will cease with prevention, and a further great social benefit is that people at very low level of consciousness will not have a negative impact on their environment. The average level of consciousness can grow dramatically by the same mechanism too. This may sound weird at first, so let me give you an example.

About ten years ago in our city, a young girl was raped and killed in the street by a man during daylight hours. It was a huge case. I was especially upset by the case because the victim was one of the best friends of my then-girlfriend. After the assailant was sentenced and sent to prison, one of my close acquaintances became his prison psychologist. I wanted to understand what kind of background could cause such a serious distortion in a man to make him able to do such horrible thing. His very deep childhood was disclosed of course; it had clearly generated the incredibly distorted spiritual structure that made the man what he was, and pushed him down to such a deep level of consciousness. But the most interesting thing for me was that this man said he was very happy to have done what he did, and if he got out of prison he would perform similar or even more serious criminal acts. He enjoyed that the world was paying attention to him, and he longed to commit even more horrible crimes to gain even more attention. Do you see what a bad advisor the ego is? It is true for everyone that we idealize what we do as being right.

But returning to this criminal, the psychological tests clearly showed that he had no conscience at all, so he was absolutely not interested in the fact that he had violated ethical norms. What's more, it definitely motivated him that the attention was on him; articles and TV news appeared about him. He experienced that terrible act as if he would have been a movie hero.

Whose fault is it that this terrible crime happened? The answer is easy: it is the fault of a faulty social system. There had been many signs that this man had severe psychological problems, but neither the people in this man's immediate surroundings nor the authorities did anything. The social system today does not deal with people's states of mind. Psychopaths can walk freely in the streets until they commit criminal acts. Of course, the problem is not that they are free as long as they do not commit any major crimes; the problem is that we do not deal with them with due diligence! With help, some of these people can be brought out of the depths that their souls got into. The rest of these people (like the man mentioned above) should be separated from society before they cause major trouble.

Such a person confronts with others several times before committing bigger crimes, keeps a lot of people in fear, radiates a strongly negative state of mind to their surroundings, and fundamentally embitters the happiness of those around them. The conclusion is that we should place our social system on a spiritual foundation. Every person's level of spiritual development must be measured, and they must be helped in reaching the next level. With this, everybody contributes actively to the development of global happiness. Unfortunately, there will be people with such a deeply damaged state of mind that they are not ready or no longer able to develop. These people must be separated from society in order not to hinder the development of their surroundings. Working out this system is a difficult task, and it differs according to the different social and religious systems. The agglomeration model is really great because each agglomeration can adapt the common principles to local circumstances. Prisons and mental hospitals already exist, but these institutions would become rehabilitation facilities where people would have the chance to heal their souls and thereby return to Life.

7.5. Kinesiological screening of important professions

It is a really huge problem in our society that people who often have an impact on others are mentally damaged. In our self-advocating world, a little spiritual distortion is needed in order to become leaders. The problem is that people who shape society should not perform such activities (with room for exceptions, of course). This could be arranged quite easily with the methods introduced in the previous chapter. People's levels of consciousness have become measurable, and this should be an obligatory screening tool in the future for every professional who has a significant impact on many others. Some of these professions include doctors, priests, teachers, politicians, influencers, YouTubers, directors, and executives of public institutions, although this list is not exhaustive.

Just imagine if the people working in these jobs were well-balanced! What a change our society would go through! These people would be an example to the world—a highly effective method which would bring significant social and economic benefits with low economic cost.

7.6. Restructuring the educational and child care system

The responsibility of education is huge. The biggest weakness of the actual educational system is that it focuses only on gaining rational knowledge. I have been teaching at a university for more than 20 years. It is shocking to see that young people between 18 and 25 have really faulty self-awareness. What most young people think about themselves has nothing to do with who they really are. If young people had real self-awareness when they left school, there would not be as many frustrated adults in the world. This is why it is crucial to make a place for effective, interactive self-awareness subjects within the educational system.

The other serious problem is that education focuses on the rational world; people are not taught about their spirit or spirituality. And at schools we learn that spiritual matters are nothing, and that we don't have to care about them. People have to be taught that there is not only a rational world, but that the outside world is much larger and it must be respected. Wisdom, feelings, and further signs of our spiritual selves can all be clues to help us beyond the borders of the rational world.

Respect for Nature and searching for happiness can only go hand in hand. Humankind has to be led back to Nature! It is not separation, but togetherness that is the solution! Nature is no enemy, but a friend! Humankind has forgotten the joy of being in Nature. It is a psychologically proven fact that time spent in Nature boosts mental balance and raises an individual's average level of happiness.

For this reason, **self-awareness, meditation, psychology and environment protection** should be embedded in education as subjects. Schools must take children into Nature more often! These things can be taught building on scientifically proven facts. We're not talking about fortune-telling by reading coffee grounds here; children have to be taught things that we can already prove scientifically.

The other main direction of the change in education, and one that corresponds to the principle of Population Program, is to have only as many children as to whom full attention can be provided by an adult during the children's awake time. The restructuring of the educational and social system to support this principle will boost the average personal level of happiness incredibly effectively! Children will be brought up getting full attention, so they will understand that they are special and unique. **The mass of children who lack love and attention will disappear.** When they become adults, these children will live with a totally different spiritual balance than previous generations.

If we restructure the educational system in a way that a teacher pays attention to only one child at a time, **children's developmental speed**

and knowledge levels will grow incredibly. This means that we have to switch to personal, individual education. Every weekend I learn English with a child for an hour, and in this way we proceed ten times faster than in the school education with its five lessons per week. Out of curiosity, I counted the difference in progress speed, and surprisingly the result was that I taught my child with 50 times more efficiency due only to the fact that a loving adult is paying 100% attention to only one child. As there are 25 children in a class on average, and the activities with communal functions must be held as well, **the individual educational system is at least twice as efficient as the traditional one.** We can also consider that it costs half as much for society.

It can be imagined in a way that the teacher spends less time with each student and gives them exercises to do at home according to the student's own progress abilities. Children with different abilities can progress at their own pace. Children with better abilities can obtain greater knowledge without suffering daily boredom from lessons tailored for slower learners. On the other hand, it is possible to progress more thoroughly with the slow learners, who always suffer from the feeling that they can never perform well. School today is hated by all but a few. My daughter detested learning English, but after we began learning it together she grew to like it so much that she completed an intermediate language exam at the age of 14, and now reads every book and watches movies in English.

Children can actually enjoy the school subjects they currently dislike. Teachers can finally enjoy their profession as being a form of talent management, so they will be more motivated as well! This system will be beneficial for every participant. **Children will get a realistic view of their abilities by the time they leave school. They will know exactly what they are talented in and in what they are not.** The number of bad career choices will fall to a fraction. Society will be more effective, and in this way there will be a reduction in emissions too, not to mention the rise in average happiness since the basis for success will be to choose a career that matches one's abilities.

Families also face great difficulties if they want to maintain the principle to always have an adult's full attention available for their child.

In order to be able to honor this principle, we have to restructure the child care system as well. It is necessary to support the parents' freedom to spend more time at home in order to increase the amount of adult attention their children receive. Grandparents and great-grandparents will play an important role in the new family models, just as they did in the old, traditional ones. This also reduces old age loneliness and depression, which are usually fueled by the feeling of being unnecessary. The personal motivation is that everybody wants their children to be happy, and it is easy to see that our child will more likely be happy in this way.

This gradual restructuring of the educational and child care systems and the support of these trends societally may seem strange within families at first, but will result in saving us socially, partly because people will have more balanced and stable lives. Addictions and dependencies will fall off dramatically, and with them all of their social and environmental negative consequences.

7.7. Missionary system among agglomerations

The missionary system of high-level agglomerations substantially boosts this developmental trend. The mission of people in agglomerations with a high level of happiness is to help those in other agglomerations, which moves development upwards. Due to this mission, people living in agglomerations with high GHIs will get achieve even higher levels. The selfless missions carried out for people living within a low GHI agglomeration are the best aid to spiritual development. The missionary system of the Agglomeration Program therefore helps to raise the levels of self-development and average happiness due to its selfless impact on others. At the same time, the living proof of benevolent missions and happy, balanced agglomerations is so positive that it gives power, enthusiasm and goals to the other agglomerations. Thus, there will be fewer and fewer agglomerations with a low GHI.

The missionary system works within the agglomeration as well. As people re-experience the joy of selflessness and its happiness-boosting effects, we will provide opportunities through the missionary system for people to experience these desires within an organized framework.

It is practical to establish missionary system coordinating bodies inside and outside the agglomeration, to which there are already several initiatives. The problem is that these currently belong to certain churches. It is worth making the notion of goodness and selflessness independent of religions. It's time to realize that every religion teaches us these, so the members of different churches should be able to work together through the missionary system. This also moves human development in the direction of the unity—and not the separation—of Humankind.

A crucial element of the missionary system is to provide the possibility of sharing good agglomeration practices in an effective way, furthermore to provide support for agglomerations experiencing difficulty (e.g., a natural disaster) and enhance peace in the world.

CHAPTER 8

The fifth program of change: the Program of Society

8.1. Future society

I introduced in the introductory part of this book that in spite of our global commercial, economic and information systems, our society still works on the basis of national arrangements. The national state system through competition between nations was the driving force of development for centuries, but it has become the obstacle to Humankind's development. It's pretty difficult to sort out Humankind's global problems and global climate change in the presence of competing national interests. Certain nations try to assert their interests at the expense of others. National security 'interests' fundamentally assume a mistrustful approach between certain nations. The unsuccessful environment and climate protection agreements made over more than 30 years demonstrate this well, as do the increasing social inequalities among the different societies on Earth. To sort out national problems, national arrangements are suitable of course. But as we will see over the next two decades, global problems will be so big that they will not be solvable by national decisions.

The Program of Society is closely related to the Program of Agglomeration. The goal is the deceasing of nation-states and the establishment of a global human society. I know this sounds scary at first. There are plenty of people who are instinctively anti-globalization, and rightfully so: they see and experience the harmful impacts of globalization every day. That's why it's important to know that **the right way is a form of globalization that takes full account of local interests.** The development built on the Program of Agglomeration solves the current Achilles' heels of globalization. Furthermore, imagine this transformation occurring slowly, gradually, over the long term!

In the transition period, there will be nations and agglomerations simultaneously. Agglomerations will receive environment and climate protection responsibilities with their connecting rights initially. Afterwards, nation-states will weaken slowly as the rational decision will clearly be the transition in the direction of agglomerations.

But now I would like to ask you to imagine a society in the far future when only agglomerations exist. The base unit of global society's establishment will be the independent agglomeration. People will live in self-sufficient agglomerations which have a local government. Local governments will delegate MPs to the **Global Parliament.** The **Global Council of Wisdom** and the **Global Scientific Panel** will deliver opinions on the work of the Global Parliament. The Global Council of Wisdom and the Global Scientific Panel will be able to accept or veto the action plans voted on by the Global Parliament. In case of a veto, they return the action plan with an explanation to the Parliament for further thought and revision. There will obviously be similar bodies alongside the independent agglomerations' local governments as well. Each independent agglomeration will have its own army and police. Countries already have these kinds of authorities to keep order; unfortunately, they will still be needed for some decades during the transition phase. As the average levels of happiness and inner spiritual balance grow, there will be less demand on these authorities. However, there will be a Global Peacekeeping Force which conducts peacekeeping and environment protection activities, and also protects the planet against outside threats (e.g., an asteroid impact). The **Global Space Exploration** Program will be connected to it. The **Global Peacekeeping Force** will consist of units delegated from the agglomerations' own forces. The maintaining of the framework directives set by the Global Parliament and their adaptation to local circumstances are the responsibilities of the independent agglomerations' local governments. It is crucial to point out that the Global Parliament never prepares concrete rules; **the Global Parliament issues directives and recognizes that these directives can be implemented in different ways by**

the agglomerations in keeping with their different cultural, economic and natural conditions. It is pointless to tell an American person to eat less beef or to tell a Japanese person that they should stop eating whale. To assert the global interest is the primary task of every independent agglomeration, but they are free to decide how to do it. Angus Forbes (2019) dreamt of the establishment of a democratic global biosphere protection organization, although as a slightly different system and establishment than the one presented here. But it at least can be seen that the central organization of planet protection has already appeared in the work of another author.

The Global Council of Wisdom expresses respect for spiritualism and life experience in society. It is not a problem if we develop slowly, but it should happen more thoroughly and carefully while maintaining and developing the greatest happiness and peace. The aim is social development while attaining perfect balance. What's the hurry? This overstrung and believingly limitless 'development' we live under can cause our final extinction. The members of the Global Council of Wisdom can only be people above a certain age who are respected in society, have an exemplary record, live according to ethic norms, have a blameless past and whose level of consciousness is at least at the level of Acceptance. The Council controls the draft directives put together by the Global Parliament from the aspect of Life protection, social balance and social happiness.

The Global Scientific Panel is the social manifestation of the respect toward sciences. This body can consist of academicians and others who have demonstrated outstanding performance in scientific work, who have lived an exemplary life ethically, and have reached the level of Reason. The Global Scientific Panel monitors and checks the draft directives of the Global Parliament from a scientific viewpoint.

8.2. What is the difference between the European Union and the union of agglomerations?

When introducing the model described in this book, I often got the feedback that there is no difference between today's unions of certain nations (e.g., the EU). However, there are two basic differences between the two social systems. The first is that current country borders do not match existing natural areas, so in their geographical locations they are not able to strengthen the natural areas. The other difference is that the European Union and the USA are the union of nations or states in which each subpart's self-interests are placed parallel to each other. This principle differs strikingly from the principle of interdependence in which the union of agglomerations exists to reach common objectives together, supporting each other.

A frequently asked question is: what guarantees are there that the agglomerations will not start competing with each other, just as countries do today? The answer is given by the comparison of the principle of interdependence, the mission system and the 'think globally, act locally' principle with the current pursuit of national interests. It should be clear that the two systems have nothing to do with each other.

8.3. The long-awaited world peace is close at hand

The gradual disappearance of nation-states will enhance the level of world peace, and Humankind's average happiness will grow with it. A war causes addictive mental imprints for several subsequent generations. The spiritual wounds caused by war are handed down from parents to children. The healing is very slow; it can take almost a hundred years. It takes several generations for a society to recover from war both socially, economically and spiritually. If there are no wars, the amount of spiritual illness will reduce significantly and social rejection will not be as prevalent either. The environmental damage and destruction of war will stop as well. Earth will breathe again, not to mention the reduction of military goods production and the resulting reduction in environmental pollution. These effects together will mitigate the widespread aggression typical in society, and also dissolve social inequality.

If we think about it, most wars break out because of the competition between nations and the fight for goods. If we do not change our way of life, the fight for natural resources will cause most of the wars in the near future. The basis for a war is that certain power groups exaggerate the differences between groups of people to extremes. The other reason can be religions whose dogmatism greatly hinders development. Although retaining the positive messages of religion is a noble and respectable human behavior, dogmatism and the existence of extreme religious movements increases contrasts between members of different religions and often interferes with social development.

Some people of course cling to the outmoded belief that other nations or members of another religion will try to crush them. This line of thinking is best avoided today; the social class who do not share those fears will grow in the future. People above a certain level of consciousness laugh at such political expressions and view with honest, understanding pity those who still fantasize due to these demagogic messages—those who live at a very low level of consciousness and need a lot of help and support to rise to higher levels. They fantasize because they receive no help and feel themselves more or better through identifying themselves with a certain social or religious group.

World peace is a realistic future if Humankind develops further gradually in the direction of the agglomeration system from the national system in the ways already shown. Society is now mature enough for world peace both technically and scientifically. The only obstacle is the social and economic system built on old habits and the Human way of thinking. The actual powers who lead nation-states or leaders with religious power are able to maintain their systems based on public fear. We know how low the level of Fear is, yet we still live with the ancient fear that we have to be scared of each other, other social groups and the forces of Nature. This is what politicians call the society of knowledge. How contrary is it? The **average person today is able to live at a much higher level of spiritualism and knowledge. This is no longer a question...**

8.4. Global innovation, research and the space program

With the progressive breaking down of nation-states, the useless overcapacity in society will fall and global efficiency will grow. For example, it will not be necessary for every nation-state to maintain a perfectly equipped research institution in every field of science. Scientific research could be merged into a larger collective unit, as could be the highly expensive space exploration programs. This way, the research centers in different parts of the world can specialize, which would galvanize the further development of science. The advance of Humankind's expansion into space could accelerate with the help of a Global Space Exploration Center, since the research and development will be done with global support. Nations and national groups will not conduct their developments at others' expense or in secret.

The gradual vanishing of nation-states will also change innovation. Technical developments are currently undertaken due to national interests (as well as business interests, which will be discussed later). Every country tries to boost its advantage over other countries by making the best use of the innovations in their countries. If there are no nation-states and the main task of every agglomeration is to make global goals their priority, every innovation can and must be freely accessible. As there will be no economic competition among the agglomerations, there will be no need to protect and hide innovations. The innovation centers in the different parts of the world will be able to specialize, and unprecedented research cooperations can be established. This will bring significant, unprecedented scientific and technical development.

Selflessness and spiritual development can develop creativity tremendously. The additional benefit of the Program of Happiness is that the large percentage of people currently functioning as little more than bio-robots with tunnel vision will become creative geniuses, resulting in more valuable ideas emerging than ever before. I recommend the Eckhart Tolle book A New Earth about the close relationship between creativity and spiritual development.

8.5. The Sun society

The notion of a Sun society is the other fundamental direction of development which enhances our society's environmental awareness and rationalizes the energy and production sector, boosting societal efficiency as a whole and contributing significantly to reaching the climate protection targets. What is the Sun society all about? At the moment, industrial production, the social system and even basic human thinking are based on the idea of being independent from Nature. Despite weather conditions and regardless of the time of day we can do what we want, how we want. Today we are prepared scientifically and technically to switch to a production system adjusted to weather conditions in most sectors. Why couldn't we produce more in a factory when the sun is shining brightly or when there is a strong wind? The energy sector, industrial production and social activities could be partly adjusted to natural processes. With this change, the energy storage demand will shrink and global society could be transferred to total renewable energy-based one even faster. Nothing else is needed other than a change in mindset, and as a result there will be a change in employee models and the automation of industrial equipment. The accuracy of our weather forecast systems is at such a high level that this can be realized. If we think about it, the basis for these processes has already begun; think about the smart electrical networks in which electricity is cheaper when there is a lot of renewable energy joining the grid.

The mutual impact of the Global Grid and the Sun society is that the development of lagging countries will speed up and basic energy security necessary for human existence will be established, and in such a way that electricity will be cheaper than ever. Furthermore, at another level of development it will be completely free.

If we adjust the Global Grid with Sun social structures, the loss of electricity transfer will be further minimized globally. The building up of a certain capacity of energy storage must be part of the Global Grid. Its size is a global planning issue and only a fraction of what the current nation-state structure wants. Only one thing is needed: the acceptance of interdependence between nations. If everyone depends on everyone, nobody will be interested in cheating others.

8.6. The eating habits of future society

Have you known that a Western European's daily water consumption is 3000 liters, of which 2700 liters are needed for the production of food (David Attenborough 2021)? The biggest portion of this goes to meat-based food production.

Have you known that the energy content of produced meat is only one-tenth of the energy content consumed by the animal (M. Berners-Lee 2019)? Earth can therefore support ten times more vegetarians than people who eat mostly meat. Future society will be totally vegetarian; meat eating will be reserved for feast days only. I know this is unbelievable for a lot people, and scary for even more. When I started meditating, after a few months I became a vegetarian, even though I hadn't intended to do so. Above a certain level of consciousness, a person becomes vegetarian anyway. A good nine out of ten people react to this idea like: 'If it means I'll have to stop eating meat, then I don't want to develop spiritually'. It may be difficult to comprehend becoming a vegetarian if through your current filter eating meat gives you joy. But as your spirit changes, your filter will change as well. People at higher levels of consciousness stop eating meat because of inner motivation. A growing number of people will be vegetarian, and it will finally be the meat eaters who are the rare and unusual ones. This is only one of the positive benefits of the Program of Happiness.

8.7. Transition to a service-based society

Maximizing personal security can cause similar overcapacities in society to the ones I mentioned regarding nation-states. Everyone wants to maximize their freedom in almost every area. As a result, everyone needs personal possession of every product and commodity. We buy and possess objects which we use only for a short term, or rarely, or even just once. This book is also like that (unless you borrowed it from a library). This causes an enormous amount of waste of materials and natural resources. Consider that if you live in a family house, you obviously have a lawnmower. You cut the grass about once a week at most, depending on the climatic conditions and your preferences. It takes you about an hour. So for a maximum usage of four hours a month you own that machine. At least other 30 families could be served by the same machine. That is to say, only one thirtieth of the current number of lawnmowers could be produced, serviced, maintained and changed when they break down. This could be done in such a way that you get the lawnmower to use when you want it. This is why it's such a powerful solution to switch from a shopping- and possession-based society to a high-level, service-based social system (Jacque Fresco 2007).

For this, it's necessary to draw up a so-called Global Service List for cases in which it is socially unnecessary to possess objects or goods. Society then has to be transformed gradually to that, by gradually forbidding the possession of the related objects and equipment. For instance, if a product which is on the list breaks down, the individual is no allowed to buy another one, or if someone doesn't have one yet, they can only use one as a service, namely to rent one. This way, the gradual transition can take place. The workforce coming from the disappearance of production-based workplaces will be taken on by the expanding service sector in the service-based society, so there's no need to worry about unemployment.

The other very big advantage of a service-based society is that we receive customized service, since the provider has the expertise to specify what we need. Think of how many times it happens in our complicated world that it turns out only after purchasing something that we didn't really need it. All useless and mistaken purchases will stop. We should also consider how many times our tastes have changed during our lifetime regarding certain things. Through services, our needs can be met without having to exchange objects for newer ones. Of course, there are things and products which are personal, so possessing them will last a long time. For example, we would never rent shoes or a toothbrush. At the same time, why would we buy vehicles or lawnmowers for the short amount of time we use them when they can be available for long-term rental as a service? With this change, production will fall to a fraction of current levels as will the impact on the environment coming from production. At the same time, our standard of living would improve; we only have to change our mindset and our consumption-based economic structure.

8.8. Labeling obligation of products and services

Most people do not deliberately make bad decisions with their shopping and habits regarding climate protection because they want to harm the Earth; they do it because they are not well informed. The capitalist, multinational commerce-industrial systems hide the information traces from us. **They hide what ethical and environmental protection rules they break to provide the given product and service.** There is still slavery, child labor and extremely difficult conditions for women in a lot of countries. Several companies produce their products in countries where they do not have to adhere to strict environment protection rules which make production more expensive. The world is supplied mainly from these countries because the workforce and production are the cheapest there. From the clothing industry to several other industries, the goods produced in poor and legally unregulated countries are sold in other solvent parts of the world. Many times, the ingredients of certain foods arrive from another part of the world to a production area from which they are distributed globally. This system that ignores human dignity and generates social tensions has to be shut down. This is the kind of globalization against which many people demonstrate. The form of globalization I recommended aims to abolish these things slowly.

In order to sort out these problems, **consumer information must be made obligatory.** If a product is a bit more expensive but was produced with a fraction of the environmental impact or without exploiting workers, people will definitely buy it. The morbid consequences of crazy price competition between multinational companies will slowly diminish from society.

Within the frame of this Program, the following labels must be made obligatory on every product: 'Carbon footprint', 'CO_2 eq value', 'Human equality ranking', 'Places of production' and 'Level of Life supportiveness'. Let's see what they mean:

- Carbon footprint label: this label shows how much raw material and energy were used to produce and transport the product, and how much its elimination will cost. This has to be converted into a carbon footprint value, whose method has to be unified according to strict rules. This allows us to decide between alternative products which is the more eco-conscious one.
- CO_2 eq value label: the quantity of emissions the production, transport and later elimination will cause it must be calculated in CO_2 eq. (CO_2 eq is all of the caused GHG values given in terms of CO_2.) People can choose between several alternative products depending on which has the least climatic impact.

- Human equality ranking label: here the fulfillment of a list including basic human rights has to be checked, and the level of fulfillment has to be marked with a percentage value.
- Places of production label: the production places of all of the ingredients and raw materials have to be recorded here, allowing a consumer to see which product was produced nearer.
- Level of Life supportiveness label: a rate has to be calculated from all of the Life-destructive and Life-supportive activities during the production, transport and elimination of the product.

The point is that all of the calculating methods of all of the labels have to be regulated and checked as to whether the companies are keeping them, because there are really serious misuses with these already in the world. For example, none of the sea protection labels on canned fish are true. If a company pays a sum of money to a foundation, they can place the label on the tin but no one checks what actually happens on the sea. In connection with this topic, I recommend the movie *Seaspiracy* to you, which you can find in the list at the end of the book.

The advantage of the labeling system is that the emissions coming from global shipping will reduce. For instance, I will be less likely to buy Chinese garlic in Hungary. A further advantage is that the market for more expensive but ethical and eco-conscious products will strengthen, so more economic power will be concentrated on their development. Another advantage is that the world will clear up in the sense that multinational companies will no longer be able to lie. Because of the requirement for straight talk, only honest companies will survive. The unethical, Life-destructive organizations will rightly fall into the economic trap if they are not able to transform.

The advantage of this system is that it will provide the basis of accounting for taxes introduced in the Program of Economy. I can't wait to get there with you!

8.9. Common world language

An important part of social transformation is choosing and accepting a **common world language.** It must be a goal that 95% of the population should be able to use this world language in 30 to 35 years. Of course, world language skills are important only in addition to maintaining our native languages as a form of cultural protection. It would be practical to choose this world language from one of the languages already spoken widely (English, Spanish, Hindu, Chinese etc.), but linguistics will make a proposal on the optimal choice. Important aspects are a simple writing system and that it should be easily learned, so an artificial language can also be a beneficial choice (like Esperanto) because it doesn't confer an advantage to any countries, and it is typically easy to learn. A world language unifies the communication among agglomerations and enhances people's feeling of togetherness. In this system, it is important to gradually experience Humankind's unity, which is one of the main keys to our further development. Highlighting our differences only generates opposition.

8.10. White is the winner - the albedo trick

One of the processes that fastens and boosts climate change is that arctic ice caps and the ice on mountains are reduced dramatically due to global warming and as a result, the average albedo of the Earth's surface will change and our atmosphere will be even warmer since the Earth's surface is able to reflect less light.

A fast and efficient solution can be that we change everything that is possible to white in 20 to 30 years—especially roads, cars, and the roofs of buildings. According to my estimations, the total surface area of these is almost the same as the reduction of the surface of melting ice, so we can mitigate the destructive effect. What is needed? Every road being resurfaced and every new road must be covered by white asphalt. I have already seen a white limestone macadam road which was not blinding even in strong sunshine. Every roof being replaced and every new roof must be white or really bright. Every car should be produced in white. I know that choosing color is also part of human freedom, but consider to what extent we would be less happy if everybody used white or really bright cars and everyone's roof was bright as well. I don't think we'd be unhappy at all. On the sides of cars various colors could be chosen: the same with the walls of houses. But the collective effect can be pretty efficient in combatting climate change!

8.11. Group meditation

Today it is a scientifically proven fact that every living creature is a special energy system, and what we see as their body is only a small part it (J. Dispenza 2020). The unity of every living creature can be easily seen on the level of energy, and most of the enlightened spiritual leaders have been trying to draw people's attention to this for thousands of years. Today's situation is a bit different because the spiritual gurus' views and scientific results are starting to approach each other; science has started to understand and prove these 'visions'. The energetic perspective of living creatures helps us to respect every living thing as equally as ourselves and our fellow humans. At the same time, this energetic perspective can help to open the world of meditation, as the existing energies can be felt this way.

The best decision of my life has been to learn and practice meditation every day, following the instructions of Joe Dispenza's book *Becoming Supernatural*. It had an incredible impact on my life and on that of my environment. This occurred a mere six months before writing these lines, so I am still quite a beginner in this field.

Dr. Joe Dispenza wrote a separate chapter in his book about the effects of group meditation, as proven by measurements and their scientific explanation. During meditation the human consciousness can affect the functioning of surrounding energy systems and therefore the decisions of surrounding people. The range of meditation is really wide; cities and even counties can be easily involved. The impacts of the levels of consciousness on others actually works on the same principle. The difference is that during a meditative state we rise to the highest level that we are capable of, so we affect our environment with the greatest good. When peace meditations have been held with larger groups, in the following days the amount of crime was reduced, or in war zones the number of deaths was reduced. These group meditations actually work that efficiently. With meditation, we not only have a positive effect on our lives, but also on our environment and on the destiny of the world!

I have seen the effects of this method on a daily basis since I became a meditation practitioner, so I can say with utter certainty that **one of the most efficient tools in the fight against climate change is regular group meditation focusing on the protection of Life.** During this meditation, people gather to concentrate together through a guided meditation for the protection of terrestrial Life, its health and expansion. I truly hope that there will be more and more groups like this in society who gather on a daily basis to meditate together for our future! I am longing to set up a group like that.

8.12. Some wrong thinking processes

The current problems derive from the same roots as in the case of the other Programs. Personal selfishness, the fear-based social system and the striving for infinite security cause the situation where the individual wants to buy and possess everything that is possible. We have to gradually switch from this way of thinking to a service-based society. The resources are not limited (unless we use them at the expense of others to satisfy our limitless selfishness) and we do not have to be scared of Nature or each other! These old habits are faulty!

The bigger part of the current obstacles is of course that a narrow segment of Humankind is still not ready spiritually to live in peace and understanding; the very idea causes fear in a lot of people, which is amplified by the lurid media.

A similar line of thought is true for the limits of nation-state systems or for the opposition of religious cultures. Why is it possible to make us believe that another group of people (a nation or religion) is dangerous to us? Most Americans believe that Russians are evil. However, both an average American and an average Russian want nothing else but to live in peace and harmony. The demagogy and incitement of powers have a negative effect on our worldview.

Not long ago, I was amused to see a short movie on the internet, in which a detailed genetic examination was done on two dozen volunteers. It was analyzed who came from which nationalities and in what percentages. It gave me a reason to smile because of how diversified the picture is, and how many different nationalities' blood flows in an average person's veins. If we look at it from a genetic perspective, we cannot really talk about nationality at all. (I am Hungarian, but there are large portions of German, Italian and Scandinavian blood in me. There is even 1% Japanese as well.) The most interesting part of this movie was when it turned out from the genetic code of an extremely nationalistic person that he had the least DNA from his own nationality—which he thought to be superior to others—and the biggest part of his genome from the nationality he hated the most! It was intriguing to see the shock on his face when he learned how groundless his exaggerated national identity was and how unjustified his sense of superiority was[1]. National "identity" is a really nice thing while it serves the protection of cultural diversity and fuels pride in it. On the other hand, national thinking at the level of world peace or social, scientific or economic systems is the barrier to Humankind's development.

It's the same with the dogmatism of religions, which tries to enhance its power based on people's fears and incites them against other religions. Just think of the medieval Christians' devastation in South America or through inquisitions, or the today's extreme Islamic terrorists who commit terrorist activities. Every world religion is basically humanitarian! Only the extreme followers of religions are not like that because they are scared of losing their power, questioning the necessity

---

[1] Many on the internet doubt the originality of this movie. I am not able to check who is right. The message, regardless of the skeptical opinions, is perfect!

of their existence in the long term. As the fear for power is anti-Human in itself, most of the churches have anti-Human features which urgently need to be weeded out if they want to survive in future society.

CHAPTER 9

The sixth Program of Change: the Program of Economy

The social torsions caused by the current economic system are also serious reasons for the current situation and make the implementation of a new system difficult. Formerly, profit interests were held back by ethics, religion and other rules. Today, everything that makes a profit is seen as acceptable, or even good. This makes society absolutely sick since it increases desire and bases everything on maximizing consumption. Buy more, experience it, possess it because you can get everything. This is only a slogan—the system is of course false! It is just a false mirage that people are chasing uselessly, resulting in nothing more than even more unhappy and greedy people. There was a video made when a big mobile phone producer launched its latest model and announced that it would be available at 8.00 a.m. in a certain store. There was a great crowd in front of the store by morning, and when the doors were opened, people crushed each other to get in sooner. Many of them fought violently with each other for a shopping basket or to enter their order. The 'cultural' structure is to push people down to a psychotic level. The profit-based 'world culture' will slowly consume everything if we do not place limits on it. This 'culture' pushes down the wonder we call a Human to the level of a primitive, consuming bio-robot.

If I am asked whether it was necessary in the past to have the current economic system, the fossil fuel-based energy, the nation-state competition, the economy-boosting effects of profit and the money world, my answer is obviously YES. These have been the driving forces of development up to now. We could not have reached such an incredible level of scientific and technical development without these. However, the things that have supported our development so far have also been the barriers to our development for some time; indeed, these systems are the gravestones of our destruction. These are such outdated fossils, just like a man sitting in an e-car today doesn't even think about his former five-liter petrol engined eco-monster—an outdated, harmful tool of mobility. The new system is about to come alive. The rudiments are already here in society; we only need to transition to the understanding that the current way is not the right way! Let's look at the tools with which we can make this transition.

9.1. GHI instead of GDP

Everybody is a part of today's money world. We cannot extract ourselves, as everything is run by money and the longing for money. It is crucial to say though that the money addiction of a spiritually balanced person is much less than that of an unbalanced person struggling with hidden or known spiritual problems. The more balanced a person is, the fewer useless things they buy, and the less they want to mitigate the gap in their spirit with bunch of useless goods. In this sense, the existence of balanced people is not in the interest of today's economy-centered social system. A balanced person doesn't buy as many things as less balanced people. A further 'problem' is that it is more difficult to make a balanced person long for things through ads and other marketing tools. But **longing is the main marketing language with which people can be dragged into being consumers.** This generation of longing is the engine of the business world. Billions of people are yearning for newer and newer services and products, and are devouring these goods blindly while Earth is slowly dying.

This is not news to you anymore. It is also obvious that GDP is not suitable as a measure of happiness or a Nature-centered society. GHI must be introduced alongside GDP, as it is a more important, overall measure as we have already seen, and one that can be measured precisely. What's most important that it overshadows the economy-centered cultural-social system. It is high time to tell the world again that Humans are much, much more than consuming bio-robots.

9.2. The companies above Humanity

Another exciting and dangerous aspect of our economy-centered social system is the exaggerated role of the companies in society that are above People. A company is said to be above People when its activities can't be stopped, even when they no longer serve Humankind's interests. These big companies exist for themselves and not for Humankind, despite the fact that they are legal personalities and do not have their own consciousness. This may sound strange at first, but I am now going to prove this statement.

In order to understand, assume that there is a global joint venture which is highly profitable but is strongly environmentally destructive and harmful to Humankind. For instance, it produces its products in Africa under carcinogenic circumstances with child labor and sells them in the developed part of the world, or it produces video games which heavily ruin children's psyches. We should admit that quite a lot of company names could be mentioned here as examples. (If you're deeply interested in these anti-Human and anti-Nature activities, I recommend Klaus Werner's *Schwarzbuch Markenfirmen*.)

Assume that the CEO of such a company is fed up with the fact that they profit from such horrible actions and tries to set the company in the right direction, which of course causes a drop in profits. At that point, the board dismisses the CEO and takes on another more cooperative CEO instead. If the board wanted to make such world-changing decisions, they would be dismissed by the shareholders, since the yield of their investments would fall. If the shareholders decided that they didn't want to invest in such a company, there would always be other shareholders willing to buy the shares because of their high yield. If the shareholders, board and CEO decided together that despite the profits they are not willing to continue the company's destructive activities, one or more competitors would immediately try to get their hands on the newly-available market and grow their production capacity.

This situation, which shows the operational problems of companies above Humanity, presents the problems of the profit-based economy well. **Profit is nothing else but the materialization of people's**

short-term self-interest. Profit is the rawest economic and social manifestation of self-interest. This is why we can't talk about selflessness where financial profit is made. We can talk about mutual benefits, or benefits gained by placing others at a disadvantage, but not about selflessness at all. The root of the problem is simple: Humankind destroys its environment and itself indirectly through selfishness since competition for money is the engine of the whole system.

The solution can be found if the labeling obligation of the Global Society Program makes businesses more open. It would soon be revealed whether the customers are bad people or the system cheats them. I am sure that the global companies today would switch to a clearer economy in no time if there was a compulsion to do so. But until this happens, we can do a lot with our individual shopping decisions and with disclosing any trickery we discover. The power of bottom-up initiatives is giant.

9.3. Profits vs a moneyless world

I believe in unconditional human selflessness, but when I was a selfish person I was convinced that it did not exist. But when it comes to money, selflessness is impossible. There are two possibilities: placing self interest in parallel, or a certain degree of exploitation of people. Capitalism is built on these two. This thought can be worked backwards: **while money rules the world, there will be no widespread selflessness.** The progressive transition away from the money world can be made in inverse proportion with the boost of selflessness.

In the far future there will be no need for money. Certain services will become self-sufficient through companies changing into professional bodies in which people do not work for money but rather for the sense of public utility. Everyone will do it, and as a result there will be enough of everything for everybody: food, water, housing, health care, education, etc. This may seem utopian from our current perspective, but if we consider how the Human spirit actually works, it is much closer to natural Human behavior than the current capitalist system with its depleting approach. Today's system is based on fear, selfishness, and the obsessive notion that goods are limited. None of these is true, and we have known it for ages. What's more, the number of people who see it clearly will grow and grow.

To understand the way out of our current situation, have a look at a far, ideal future where human society works differently. In the distant future, there is no money and no profits, but there are still companies, no matter how strange this may seem. To be more precise, they are no longer companies per se, but professional bodies. Humankind lives in technologically modern cities in which motivation is the driving force. Every spiritually-balanced person's dream is to live and create in peace. The desire of people with normal mindsets is to do things which benefit other people or Nature. If all the material needs of a person are provided and they don't have to worry about uncertainty in the future, they are able to live in this way. Of course, to reach this point the healing of spiritual wounds is crucial. A spirit loaded with addictions is not able to take in good without conditions; it always generates some unrest around itself. But let's assume that people in this distant future don't have to struggle with deep spiritual problems because they have already healed slowly through the generations. In such a world, people work out of devotion and selflessness. They are driven by the desire to do things which serve other people. As everyone lives and thinks alike, everyone can do what they like. Everyone works flexible hours and the system works perfectly. The basis of the whole system is that everyone feels secure, so they do not need profits. People do not have to fight with each other or Nature to get more, because everyone has enough of everything. They don't even have unrealistic desires because the lack of spiritual wounds means that there are none of the exaggerated human needs which seem so common nowadays. A spiritually balanced person does not long for unlimited power, wealth, a perfect appearance, etc.

The basis for profit and the power gained by it is the inner fear that we can find ourselves in trouble at any time. But if you consider it, we have nothing or nobody to be scared of. Unfortunately, this doesn't mean that there aren't parts of the world where people have to live in fear. It means that we are able both scientifically and technically to live in a safe, peaceful and happy world where there is no fear. The principle of evolution, based on the levels of consciousness, says that a species' next level of development is exactly what enables them to give up this old-fashioned thinking process. To achieve this, we have to gradually stop those economic habits which are based on the old images of fear and selfishness.

You might recall that this book began by sharing the fear of our realistically possible future regarding climate change. However, the fear-based system is the cause of the climate crisis and many other social crises, so giving up fear-based thinking processes and the transition away from fear-based social systems brings the world to where it can solve the climate crisis and set Humankind on a long-term development process.

How can we switch from the current profit-oriented economic system to the motivation-based social system introduced earlier in the mid-term? Gradually, through many small steps, and working across generations. Let's have a look at the main tasks and motivations of this work.

9.4. Deleting the principle of limitless growth

The modern economy is built on the principle of limitless growth. It probably doesn't need to be explained that this is not a realistic principle, or at least until Humankind starts to populate space. In order that our development reaches that level, we have to learn first to set human society's development on a long-term path of balance. **The principle of limitless growth must be deleted from today's modern economics and replaced with the principle of striving for the perfect balance.** This means that development should happen only in a direction which brings greater balance than the previous state. Balance and infinite creativity bring the development of the future, not the principle of limitless growth. This reorganizes economics from its core, although is not too difficult to change it in this way. To provide details about the transition would require a whole book, as it would be necessary to analyze it in detail on the levels of microeconomics and macroeconomics. The page limit of this book does not make it possible to include that analysis here. As agglomerations strive for uniqueness and the best adaptation to local cultural and environmental circumstances, I can easily imagine different economic preparations of the principles with the condition that the main objectives are the same.

9.5. Tax relief and interest rebate

This is a topic I am going to talk about shortly since there are big breakthroughs in this field. The point of the system is that those investments, services and products which are greener and more climate-friendly must be given tax relief in certain countries. In the case of investments, the interest rate of bank loans will be reduced according to how climate-friendly the investment is. This set of tools will be needed in the transition period to speed up the process of change.

9.6. Taxing of derelict land

Taxes are also the tools of the transition period. While there is money, there will be taxes.

We spoke about Area Usage Ratio in the Program of Agglomerations. This shows how many redundant areas are maintained by an agglomeration that could be given to Nature. It also shows how we waste our lands. This tax must be introduced in terms of the social expectancies of efficiency and the support of Nature's expansion. This tax must be introduced on every piece of land which is located inside the agglomeration and is unused, i.e., it lies fallow. In order to avoid system misuse by false usage, the detailed rules will have to be worked out of course.

Unfortunately, there are many lands lying fallow for investment purposes nowadays, enhancing the vegetation and habitat fragmentation, while these lands could be useful parts of the agglomeration or the network of natural areas. Taxation will force speculators to sell these areas quickly to people who really want to make use of the land. If the land is cannot be sold for a period of time, the agglomeration will have to take it from the owner in return for fair compensation, and this property will form a part of the area change process, which aims at expanding natural areas.

This tax aims to reduce the agglomeration's area and give greater space to the surrounding natural areas without a drop in the standard of care.

9.7. Environmental Impact-, People Harming- and Labelling Taxes

I have already mentioned in the previous chapters that I would talk about the funds needed to start the changes. The previous subsection was about this, and this one is too.

The Environmental Impact Tax, the People Harming Tax and the Labelling Tax must be introduced. These taxes must be applied on all the products and services which harm either Nature or People. The income from the taxes must be spent only on environment protection and spiritual development. The agglomerations' local governments decide on this income. After the system is globally completed and the Global Parliament is established, agglomerations pool some amount of the tax income. The Global Parliament decides on the principles of distribution, involving the Global Council of Wisdom and the Global Scientific Panel.

It is worth looking into the two mentioned taxes to understand their socially beneficial effect. Let's start with the **Environmental Impact Tax.** This tax must be paid on two things. One is the mining of any raw material or exploiting any natural resources. For example, if we mine coal, cut down trees or use soil, we have to pay taxes. With this tax, **we pay the cost of the things taken from Nature** and from this money we give back to Nature in another way. As Nature has rights, we cannot deplete it freely. It's strange that currently we can take anything from Nature for free, isn't it? The other **tax comes from the environment loading caused by shipping between agglomerations.** If I want to buy South American aloe vera extract while living in a Hungarian agglomeration, I can do it, but the product will be levied with the Environmental Impact Tax. So as a customer I have two choices: if I do not want to pay a lot, I can buy local alternative products. But if I insist on buying the product, I have to pay a higher price charged with the tax. In the case of products coming from outside of the agglomeration, the level of the tax depends on how much environmental loading the shipping causes. For instance, if the company works with a transportation company which uses only solar powered boats, this product will either not be taxed (in the further future) or will be taxed much less (in the near future). Tax will have to be paid by both the producer and the customer for there to be incentive to lead society in an environment-conscious direction. One of them will be motivated to change to more ecological shipping options, and the other will buy remote products only if absolutely necessary.

You may ask why the products produced within the agglomeration are not charged with the Environmental Impact Tax, since it can happen that the production of a locally-produced product is actually more harmful to the environment than a remote product and its transportation. The reason is that the agglomeration is responsible for balancing the environmental effects of the activities within the agglomeration and Nature. The way the agglomeration achieves this balance from the balance strategies already introduced is its own internal affair. For instance, if the agglomeration taxes a certain production process, the tax will be the direct income of the agglomeration, not a global tax. But it can also happen that the agglomeration will accomplish the switch to a more eco-conscious production process through economic incentives or by changing the legal regulations, and not through taxes. It can also happen that a given agglomeration is surrounded by such a large natural environment that it doesn't have to do anything as the agglomeration is already in balance. There will surely be a lot of agglomerations in place after setting the borders of the agglomerations globally and determining which of their emission balances already meet the requirements for balance with Nature. In these agglomerations, people 'only' have to work on the development of the GHI. Now let's move on to the **People Harming Tax**. This tax is even more interesting than the previous one. It must be levied on all the activities and products which are harmful to human health or spirits. Alcohol, drugs and tobacco products must be taxed, as well as all the products or services which are marketed through selfishness or spirit-destructive ads. This tax must be levied on pornography or horror movies, or on video games containing aggression or causing addiction. It is not about it being forbidden to watch porn videos, play video games or smoke. These are all possible, but since they all damage the human body and/or spirit, the producer should pay the tax and at the same time the customer should pay it as well. The income from this tax must be used only for improving people's physical and mental health. With this tax, customers will support the activities that develop the GHI, and at the same time the producer will be interested in producing less harmful products or taking greater care with the ethical content of their ads. This single tax will fundamentally alter the development direction of the economy if it is significant enough to be taken seriously by companies.

The Labelling Tax will function on the basis of the label values introduced in the Society Program. Those companies whose label values and technical principles are among the best will not be taxed. The other companies are charged taxes such that the more their products differ from their label values, the more tax they will have to pay. The level of taxation differs in the different agglomerations as the global environmental load of a product produced on site is likely to be lower than a product produced in another part of the world. In other words, the same product or service may have different label values depending on the agglomeration in which it was made, thus mitigating the unrealistic global traffic and the unreasonable global market structure.

There are overlaps between the three taxes which must be managed through the clarification of the rules. There are several opportunities to do this, but there is no point going into details in this book since a solution can be reached in many ways. Obviously, it is true of all three taxes that they will have to be introduced gradually to allow time for the economic transition.

9.8. Mitigating economic inequalities through agglomeration rules

In connection with agglomerations and local incentives the question arises of what happens if there is a giant producer in the agglomeration, delivering to every part of the world (e.g., a car manufacturer). How can the agglomeration reach balance with Nature in this case?

The answer is simple: we have to consider only the emissions made directly within the agglomeration. If this company generates excess emissions which cannot be reduced with technical intervention to the required level within a realistic time frame, the company has to relocate some parts of its production to other agglomerations. This process helps to create new job opportunities in places where there is unemployment and to reduce production where there is a labor shortage. Capital will move to the underdeveloped parts of the world. On the other hand, if the company divides its activities among several smaller factories, products will arrive at the assembly plant from several other agglomerations. These would be charged with the Global Environmental Impact Tax. But in this case the tax is also lower if the company transports products across shorter distances or by transportation methods which have less environmental impact. This process generates not only the expansion of capital but gives a boost to the development of ecological ways of transport and balances the economic differences of societies, thereby reducing social tensions as well.

9.9. Solidarity Tax - the principle of direct and indirect emissions (breaking the lobby power of environment-damaging industries)

Direct and indirect emissions can be defined by each agglomeration. Direct emissions are those made within the agglomeration. Indirect emissions are those which are generated by products arriving from outside the agglomeration due to its consumption. The agglomeration is responsible only for its direct emissions, which is to say that it has to provide balance with the surrounding Nature only with respect to its direct emissions.

However, it is the moral duty of the agglomeration to work actively toward the reduction of its indirect emissions. The best way to do this is targeted information and education affecting consumer habits, as explained in the Program of Agglomeration. At the same time, an economic tool can be the so-called **Solidarity Tax.** The point of this tax is that if an agglomeration wants to help with the problems of another agglomeration, it can do so with this tax. Here's a simple example. Let's say we know that overfishing done by European giant boats along the African coasts causes really serious food problems and threatens to cause a huge ecological disaster. A European agglomeration could decide to charge seafood products with the Solidarity Tax. Half of the income from this tax goes to the agglomeration which charged the tax, while it transfers the other half to the agglomerations which suffer from the problem. With this process, agglomerations can help each other and a cooperative network can slowly develop which is resistant to the Nature-destructive lobby powers and enhances peace in the world.

9.10. CO_2 extraction from the atmosphere

I have already mentioned CCU/CCS technologies, the development of which inevitably belongs to our generation's most urgent tasks. The reason is that it is a scientifically proven fact today that it will not be enough to become carbon neutral globally by 2050. To avoid climate disaster, we also have to extract a lot of the CO_2 from the atmosphere that got there artificially.

To speed up these developments, introducing a carbon tax is necessary above a certain company size. Of course, there are already social rudiments and a background for this, but it needs to be institutionalized and more widespread. Those companies which extract their own direct and indirect CO_2 emissions from the atmosphere can be exempted from the tax. This also helps in boosting these developments and provides a market for these technologies. The companies which do not engage in such activities have to pay the tax on 120 to 150% of their emissions. It is necessary to pay for more than their direct emissions because the extraction of past CO_2 emissions and the extraction of the company's indirect emissions have to be financed as well. The tax income can only be used for other CO_2 extraction from the atmosphere or for developing such technologies.

9.11. Switching to regenerative agriculture

The result of an increasing population is that more and more agricultural land needs to be broken up for housing. This process is accelerated up by the practice of agriculture abandoning depleted lands after the land is infertile and moving on. If we go on like this, Humankind will be using more than 80% of Earth's natural resources by 2050. This would result in the total collapse of natural eco-systems, so it is in our primary interest to slow or stop the expansion of agriculture. This what the Program of Revitalization and the Program of Agglomeration expect from us. As the border of the agglomeration will be given, agriculture will be forced to no longer destroy soil since the agricultural lands will need to be used indefinitely. Regenerative agriculture is a solution; its basic principle is to keep the soil in good condition. Furthermore, the exploited agricultural lands which are in bad condition can be rejuvenated with the help of these agricultural methods. I am not going into detail here since there is a lot of literature available, like Paul Hawken's *Drawdown*. You can find the names of two good films about this topic at the end of this book.

9.12. The Global Service List and economic restructuring

I have already discussed the Global Service List, so here I just want to add some thoughts in connection with the economy. As we know, there are services in this list in connection with products that can't be owned in the future, and that from a certain time can only be rented as services.

The list must be expanded gradually and the organizational basis must be established for the changeover during the development of the agglomerations. The companies that gained huge specific expertise in the past from production can compensate for the drop in income coming from low sales when transitioning to professional services and a rental portfolio. The future of these companies during the transition is assured in this way. They will transform into companies which produce, provide service assistance for, and rent out their products. A truly professional and high-level service can be established this way. The result of this social transformation will be that the aim of product development will be to maximize product life and reparability, as the service will be the most profitable that way. If products become outdated in the short term, the company will not be able to rent them out at a reasonable price. The direction of development will therefore change to durability and reparability.

This will result in having a fraction of the former waste, and the useless material and energy consumption will definitely fall. The reason on the one hand is that overbuying will disappear in the given sector, and on the other hand because not everybody will possess the product, much fewer products will be produced for the same service level. Obviously, the saving proportions are different depending on the product type, as there are some products that are used by many people at the same time while others are used only rarely. Furthermore, there are products with short amortization times, while this can be expanded significantly with other products.

What is more exciting is that the public will benefit as well! People will receive specific help to always rent the product that meets their needs. Let's look at a simple example. You won't have to buy a car and use it for ten years. You will be able to drive a different type each week as you please or as your current traveling needs require. If I go on holiday with the family I will get a van, but if I'm just getting around in the city alone, I can get a small e-car. If I go on a long business trip, I will get a long-range e-car or—if it's too slow—a diesel car.

Since everything is reparable, there is a need for a wide range of services which can give people interesting jobs that suit their needs and abilities. This will also solve the problem of worrying about unemployment. The same number of jobs will be needed, but they will be of higher quality and focused more on creation. We will not be the service staff for robots; there will be no need for profit-oriented forced developments or profit-oriented material savings during product development. Customers will no longer be cheated by these. The notion of planned obsolescence—one of the biggest ecologically self-destructive elements in the current economy—will disappear from society.

When services have become dominant, owning items will be a nuisance and a problem. Ownership means obligation; it doesn't offer variety or freedom of choice. We don't even recognize how bound we are by owning a house. This is true for every form of possession. **We give up our freedom for our sense of security. The freedom resulting from this new system boosts the happiness of people.** Just imagine if everything you owned fit into two suitcases, and you rented everything else. Such enormous freedom and mobility it would give you, and how much easier your life would be! What is needed to accomplish this? Only the abandonment of the aforementioned fear-based way of thinking. If you want to possess everything because you are afraid of your fellow Humans, your chances for happiness will be dramatically reduced. This may all sound weird at first, but it is true. It is also true however, that the transition cannot happen overnight. To become open to each other can happen only gradually in a selfish society.

This development can be enhanced even further. Imagine when the world starts to function without money. Those who still have possessions will not find anyone to repair, clean or rebuild them, simply because the selfless public social systems will not work with possessions. That is to say, possessions will be nuisance and a problem.

This process will start with more and more very wealthy people giving away more and more stuff to others, which will help improve the distribution of wealth in the world. The joy of selfless giving must be built into the education of younger generations, thereby helping a lot to speed up this process.

9.13. The gradual disappearance of competition

In the Program of Economy, I should talk about the issue of competition already argued in Chapter 2. We now know that the main driving force of evolution is not competition but cooperation. So, classic economics based on the evolution model of Darwin has failed as well. The competition-based economic system can only be a Life-destructive one. The processes of change introduced so far will result in the weakening and demise of competition. Agglomerations will not compete with each other because the principle of interdependence will reduce competition significantly in the economic sector. Giving up money will mean the final disappearance of competition in the far future, which also generates the realization of world peace.

9.14. Circular economy, zero-waste future

A profit-based society will never be able to produce little waste since the basic concept of the social system inhibits it. However, the switch to a service-based society enables the establishment of a circular economy—which has long been discussed—and the almost zero-waste society. Interchangeability will again become an important part of today's economic system with its disposability approach. The throwaway concept increases profits since the producer does not take responsibility for the product all through its life. Since we have to pay tax for the goods taken from Nature, it will be more profitable to use recycled raw materials than natural resources. A good example is this book, which became more expensive since it was made from recycled paper. How ridiculous this is!

These processes will bring our economy closer to the ideal model of a circular economy. The closer we get to it, the greater the chance we have of being in balance with Nature.

CHAPTER 10

The interaction of the 6 Programs; transition to another direction of development

All of the 6 Programs were created in a way to turn Humankind gradually in the new direction, starting from the actual social system. What's better is that none of the 6 Programs function in isolation, but rather they are effective while strengthening each other. Imagine that human society is descending on a downward spiral. **These 6 Programs** stop this process at first while strengthening each other, then through complementing each other's effects they **set Humankind's development on a rising track.** I'm not saying that the 6 Programs are perfectly worked out or infallible. In this book I have introduced their concepts. If more and more People accept these directions, several directives have to be set—involving scientists, professionals and creative thinkers—which can start the actual implementation after accepting the directives globally or introducing them gradually in progressively larger regions. It is high time to act and change, but doing so demands proper objectives. I hope this book helps with setting the right ones.

At the same time, the 6 Programs help us think about how wrongly we are living and how much more wonderful a life we could provide to our children if we turned ourselves in that direction. **Don't forget that our generation's task is the urgent transition to another direction, so please do whatever you can!**

One common positive element in the different Programs, especially the Revitalization and Agglomeration Programs, is that **they can be adapted at the level of the individual, family, community, country or group of countries.** So, in this program scheme, individuals, families, politicians, managers, and leaders of communities and countries can find their place and add what they can or would like to. If we consider this issue with regard to the Revitalization and Agglomeration Programs, the individual can plant trees regularly, take part in public tree planting or organize such events to reduce their carbon footprint. However, there are not enough tree plantings, they are held in an ad hoc, disorganized way, so this activity must be improved. But if we consider this issue from the perspective of the Revitalization and Agglomeration Programs, it would be reasonable to organize these tree planting events in places where the existing and relatively unharmed areas of the ecological network are expanding. Namely, these plantings should be organized at the borders of natural areas, nature reserves and on the narrowing parts of these areas in order not to combat CO_2 emissions only, but also to try to support the damaged, degraded, and weak ecological systems. If we plant 2000 trees during a public event at the edge of town in an area surrounded by agricultural areas which are worthless ecologically, it's a really valuable undertaking because those trees will extract CO_2 from the atmosphere and we are happy about it. But we also have to make habitat, and that kind of habitat will be worthless; from an ecological perspective it is similar to a desert. On the other hand, if we plant those 2000 trees next to a biodiverse, boggy, wooded reserve, they will integrate into this ecological system and the habitat of rare species there will grow. We can thereby reduce the loss of diversity coming from habitat and vegetation fragmentation. What's more, through increasing the ecological network we can help the migration and expansion of these species and the strengthening of the whole ecological system along with it. Therefore, the Revitalization and Agglomeration Programs together assist the few individual ecological and environmental ambitions in coming true as efficiently as possible.

  The other interesting aspect is that politicians and city leaders have surely already realized how little money and how few opportunities they have available for environmental and climate protection. The abilities and financial possibilities of an average municipality are miniscule compared to what they would really need for climate protection. On the other hand, if we think in the context of the agglomeration system, the leaders of the settlements can cooperate and make a broader community according to the principles of the Agglomeration Program, and the resources could be used systematically. That way, the combined resources end up in the best places and make the best investments. The resulting social efficiency is much higher and the protection of ecological systems happens in the most efficient way as well. In addition, the whole agglomeration as a community gains the most benefit in the mid-term from the environmental, ecological and climate protection developments conducted in tandem. The demarcation of an agglomeration is based on the demarcation of the surrounding ecological systems, so it's not certain that there will be valuable ecological systems on or within a certain settlement's administrative boundaries. There may be two settlements close to each other such that their pooled climate protection investments are far more efficient than the same investment made within a single settlement. At the agglomeration level, if a two, five, or ten times larger ecological investment is carried out elsewhere for the same price, it does not cause the disadvantages to the settlement that the same investment would if carried out within its administrative boundaries. Therefore, this program system can also be well applied at the level of cities and other communities.

The situation is the same at the country level if the leaders divide their countries into ecological agglomerations and structure the programs of climate and environment protection investments according to those. In this situation, money will be spent more efficiently and the results will come much sooner. Agreements can start between countries as well, since borders in most cases were not set according to the boundaries of valuable ecological systems. A very good example is the Water Framework Directive of the EU, which lays down that when there is a river catchment area that crosses national borders, the countries have to cooperate in improving the quality of the waters and lakes in that area. If we follow the same way of thinking at the ecological agglomeration level, a cross-border agglomeration can be established in order to efficiently protect the shared ecological system. International agreements can be made, and the Programs of Agglomeration and Revitalization can thereby be adapted well on the level of groups of countries. These activities at different levels bring the structured establishment of the agglomeration and revitalization fields, which will result in the opportunity for all the other Programs to become agglomeration-based.

It is also crucial to mention that another big advantage of the agglomeration system and the connected Programs is that several current environmental regulations can be mitigated within the agglomeration as the conditions must be provided at the boundaries of the natural areas. Today's exaggerated environmental regulations can make communities' development difficult, which is a barrier to development. On the other hand, if the development of human society is the aim of the agglomeration and the respect and protection of Nature is undertaken in natural areas, certain regulations can be mitigated there. For example, there is a regulation in Hungary at the community level that the proportion of green areas within the settlement cannot be changed. However, these regulations can be partially lifted if the settlement is surrounded by natural areas. These kinds of mitigations in the regulations can help a region to reach its development objectives.

I would like share with you one more line of thought about the environmental and flexibility benefits within an agglomeration. If the natural areas surrounding the agglomeration have rights, and within the agglomeration people have the priority rights, then people living in the agglomeration will only have the task of meeting the environmental targets at the boundary between the agglomeration and the natural areas that provide the balance between them. Let's look at a simple example of why this provides several advantages compared to current environmental regulations. Today, if we consider the Water Framework Directive, apart from seriously altered water bodies most of the water bodies must be brought back to a natural state, and the aim is to restore every body of water to good condition. This is a really serious environmental target and will cost a lot of money for all of society to work out. If we do not have to bring the waters to a semi-natural state within the agglomeration since we can think at the agglomeration level, we will have to provide only the water quality conditions which correspond to the assimilation capabilities of the natural areas only where waters reach the boundary between the agglomeration and its surrounding natural areas. In this case, the reduction of water-polluting emissions in the agglomeration has to be made to the extent that—considering the self-cleaning capacity as well—the two together can maintain the limiting value. This is a huge relief and a significant investment reduction which can be spent on the release and revitalization of natural areas. We can think the same way within an agglomeration regarding air emissions, waste management, and in many other environmental fields.

CHAPTER 11

The individual's role

It's not possible to give a complete and full list of the things an individual can do to implement the targets of the Programs, since everyone has a different lifestyle and different opportunities. You can find continuously growing information on our website and on my blog about the topic, and we also starting advanced training in special fields so you can go deeper in the practical adaptation of the things you have read here.

It's crucial to highlight that **creativity and unique ideas** are always the most valuable at the individual level. This chapter is a sort of thought-provoking summary about what the individual can do in their life to live in a nicer and more ecological future and to be able to avoid the distress caused by climate change. I would like to ask you to feel free to choose from this chapter and implement whatever you can in your life.! My other request is for you to expand this list with your creativity and send me please your ideas to the e-mail address found at the beginning of the book because I am writing another book—a practical guide for individuals—and I will be happy to add your ideas to that book. That way, you can be part of the act of saving the Earth. **I believe most in the power of community and in individual creativity, so I believe most in You, dear Reader!** But let's see what I can offer you as a non-exhaustive list!

11.1. The four main principles of carbon neutral life

Today it is possible to live in a totally carbon neutral way in balance with our environment in both a scientific and a technical sense without giving up any kind of convenience related to an average person's needs.

If you want to reach this, you can do it with four conscious steps:
1. Become aware that it's possible, and learn about the technical, financial and other possibilities.
2. Change your old thinking processes and habits.
3. Prepare a climate balance.
4. Start to sacrifice some money, work and time to become carbon neutral step-by-step, gradually.

The preceding chapters have been about the first step so far. Paul Hawken's Drawdown provides a very good summary about the technical opportunities, but you can find plenty of sources on the topic on the internet. You don't have to know everything yourself; here are lots of professional companies, experts, and designers good at special fields in the society who help you if you have the right gal.

This book is about the second step and the following chapters as well, and I hope it will be helpful.

Climate balance means taking into account all the GHG emissions of your family or company, and check how much emission is caused by certain activities. A lot of websites will give you professional help with this. Furthermore, there are companies who can calculate it for you if you don't have the capacity. We are happy to do it as well, so if you need help, you can contact us with confidence!

Calculating one's climate balance is good for seeing how much GHG emission you are responsible for. Better still, you can see how the activities are compared to each other, allowing you to start in a targeted way with those activities which are the most harmful to the Earth. As environmental consciousness is an expensive thing, you may only be able to proceed with the developments step-by-step. However, a sequence is going to be created for that. Since we're talking about money, it's important to highlight that there are more and more public and corporate tenders on climate protection investments, so it is worth trying to save some money by starting with these tenders.

The fourth step is simple from here, but tough as well. It is simple because you have exact goals and ideas about scheduling, but tough because we have to make sacrifices to realize them. I started these developments about ten years ago and have reached that my family's GHG emission has been reduced by more than 70%. I've set a goal to become total carbon neutral by 2030. This obviously has its price. We do not go on holiday to the Bahamas for example, but to closer places. We'd rather spend that money on solar panel development or making the payments on a e-car. Our family is happier about it, our conscience is clear, and we feel that we are helping the future with our selflessness!

11.2. The power of an individual's decisions and example, or the principle of impact multiplication

There is a Hungarian saying that "you should sweep in front of your own house first". It means the ame as the English expression that says people who live in glass houses shouldn't throw stones. Everything that is needed for guidance is in this saying. Your personal decisions can speed up and also slow down climate change. The root of almost 8 billion individuals' actions shape the world. **You cannot afford the luxury to remove yourself** and say: 'what I do is just a drop in the ocean and is worth nothing' because it isn't true! By sweeping your life clean through your actions, your conscience will be clearer and you will be happier! If you don't want to do it for the fight against climate change, do it for yourself! If you don't want to do it for yourself, do it for your children! If you don'tt have children, do it for your loved ones, or out of respect for Life!

The individual's role is giant, no matter how little it seems. Acting somehow does not only form our lives, but our example impacts others. Our children, families, friends, colleagues and relatives learn about us and the change will also impact them to different extents. Furthermore, our decisions have an impact on the sales habits, production and services of companies. Humankind works like 8 billion connected cogwheels. If one cogwheel starts to spin slower or changes direction, it has an impact on the motion of the other connected cogwheels. Those cogwheels affect further others while the whole system starts a new spectrum of movements! **You are one of these cogwheels, so please start to spin differently!**

What's more exciting that there have never been so many opportunities in Humankind's history for individuals to impact others, meaning that **we are able to multiply our impact** using easy methods. The opportunities provided by the internet and social sites are limitless. These systems are used for so many bad and harmful things. Why couldn't they be used for good things with the same efficiency? So while you're sweeping around your house, share your experiences with others and inform us as well! At the e-mail address found at the beginning of the book, we are happy to read any good news about what you've achieved in the common fight for saving the world!

11.3. Your personal happiness is the most important

As you already know, a Human's first and most important task is to do everything for their happiness! **A happy and balanced person is the most environment-conscious person and the least dangerous to others.** However, if you expect to be happy because of having a new car, a bigger house, more businesses or gaining a doctoral degree, unfortunately you are on the wrong track! Happiness comes from inner spiritual peace. If you find it, you will reach your goal. So first sweep around your house! Be happier! Live in greater peace with yourself, live in harmony with yourself, find your spiritual balance, and you will make peace with the world as well. Once you have reached it, you will live a less Life-destructive existence! How you can get there is the topic of the other book in this series, but you can find a lot of spiritual literature and help on the topic. It's no accident that this topic is so popular today. People feel this is the right way, even though they are a little uncertain which way to go due to inexperience. Once you have reached happiness, start to do the same for others. At first, obviously, work towards the happiness of your loved ones and friends. If there's no point 'sweeping' there any more, do it for the other people or for the other living creatures in your environment!

11.4. Everything that boosts your selflessness saves the world

The more you do for others, the closer you get to your deeper inner peace and spiritual harmony! This book is also made from total selflessness. It has no other aim but to help and that's why I feel so good. Creating boosts energy, and selflessness makes everyone incredibly energetic. However, only those who have reached the level of consciousness where real selflessness exists can truly feel it. Unfortunately, people generally don't believe in this until they've reached a certain level of consciousness. But I know it does exist, as I have experienced both sides.

A Human is basically a social creature, but selfishness makes us forget about it slowly. We don't even realize that this is why we're lonely and why something feels wrong inside. There are limitless possibilities in community activities, so try to take part in more of them, or organize one yourself! We were with my family and colleagues at a public tree-planting a few days ago. It was a very happy, pleasant public experience that had a positive impact on our spirits, not to mention that we worked against climate change by planting hundreds of trees. Everybody took part voluntarily in the event and everybody was smiling and happy. Only goodwill was needed, but it was really efficient nonetheless!

So take part in climate protection demonstrations, public garbage collection, public tree planting, peace mediations ... the list is infinite! Don't be indifferent: be part of the community. I would like to gather an active community on our website, where people who want to do something for the world can get together. We try to organize a lot of these events with them. Come and visit our website—we will be happy to see you there!

11.5. There is a huge potential in personal Lifestyle change

When I started to deal in depth with the impacts of personal lifestyle changes on climate, I was also surprised how much potential there is in this issue. This chapter cannot be complete; instead I've looked into some topics just to encourage you. If you read them, I hope you feel like expanding these ideas further in your own life!

11.5.1. You can save the Earth with your eating habits!

The two most terribly environment-destructive kinds of food are beef and palm oil. If every person stopped eating them, it would save the Earth. What's more, there would be no starvation for Humankind. It sounds unbelievable, but it's true. Let's see why.

Palm oil can be found in most foods, since it is the cheapest edible oil. This is why multinational food companies use the most in their products. We pay a very high environmental price for it, as most rainforests that are cut down become palm plantations. We are burning both ends of the candle with this. In burning the rainforests we emit a huge amount of CO_2 into the atmosphere, and at the same time we destroy the ecosystem which would otherwise provide the maintenance of climate balance by binding CO_2. How can palm oil use be reduced? This is a difficult question because most non-perishable foods already contain it, and surely you don't have the time and energy to read all the small print on every product. Producers have even given several different names to it to hide it from customers. The easiest way is to buy locally produced products instead of multinational food industry products. There is rarely palm oil in locally produced products. Doing so helps the environment in an additional way as well, since the transport of local products causes less impact on the environment. Of course, it can also be a solution that you read the small print and start to weed out food from your life that contains palm oil.

Beef production is responsible for the other half of deforestation. Beef is more harmful than palm oil since it does not cause only the destruction of rainforests, but cattle release a huge amount of methane into the atmosphere during their life. To produce 1 kg of beef takes 70 times more land than growing 1 kg of fruit or vegetables. So if we didn't eat beef and grew fruit and vegetables instead in those areas, there would be no starving on Earth. I know starvation depends on many other factors as well, but this statistic clearly shows how harmful a social trend it is to eat beef. One beef hamburger is responsible for the same amount of CO_2 emission as an air conditioner running for 24 hours, or driving 30 to 40 km in a typical car.

If you change from beef to chicken, only one-fifth of the land is needed to grow fodder and its GHG emission will be only one-tenth! You will still be a meat eater, but you would do a lot for the environment!

You can see that being a vegetarian is an ecological thing. But I'm not asking you to bea vegetarian if it's impossible for you. I became one. Many people are vegetarians not because they want to save the Earth but because they feel better this way. It's the same with me, but for ages I was a meat eater. I loved eating meat and didn't want to give it up. But five or six years ago I started to reduce the amount of meat in my diet. My body was grateful because I felt better. At the same time, I saved the world significantly with the change and it was a good feeling for my spirit. In this way, I did good to myself on several platforms. I'm not saying that I no longer eat beef at all, just that I eat it very rarely. On special feast days, once in two or three years I'll eat a good steak or cook a beef stew. But while I used to eat a lot beef, after I learned about these things I limited it to special occasions. My standard of living hasn't fallen at all, but rather it improved from certain aspects.

11.5.2. Travel closer. Fly only if it is absolutely necessary!

I know a lot people who are keen on traveling. It has become a significant leisure time activity for western civilization. It's found to be the best way to spend the redundant time and money produced in places where people live better economically. Traveling is really great: it's amazing to discover new landscapes, countries, and cultures, and the change of scenery reduces stress. To experience the magic of novel things caused by new experiences makes us forget about our everyday spiritual problems. Due to these reasons, anyone who can afford to travels. Tourism has grown incredibly, and the transport and industry built on it have a significant impact on climate change.

You already know that your part of the CO_2 emissions of a Budapest-New York flight is the same amount as the Earth's CO_2 binding capability of a person during their entire your lifetime. This means that if I travel to New York, I generate as much CO_2 as my part of the CO_2 bound by the Earth's ecosystem. I have always known that flying is absolutely polluting but I never knew it was quite that bad. I am not a travel addict, but if I add the flying distances of some international conferences and youthful traveling, I have already flown the Budapest-New York distance at least 12 times in my life. Since discovering the fact mentioned above, I have had a bad feeling about the flights in my life. I have already used up all the CO_2 emissions for the next 12 generations, and my other energy-consuming activities are over and above this amount! Because of this, I've thought about how many flights would have been indispensable. Unfortunately, looking back: none of them were! I could have saved it all for Earth if I had been aware of the weight of my deeds. Furthermore, I can confidently say that I would not be any less happy if I had skipped all those flights. It's true that I would have had fewer experiences from faraway countries, but I'm pretty sure there would have been other positive things to replace them.

I know several people who have already been to more than 50 countries which you can reach only by air. Traveling has become their life. If we consider it from their point of view, it is a wonderful thing. But if we consider that they have already burnt enough CO_2 in their lives for about 100 generations, it is not so nice. I am not blaming them, because they are not aware of the weight of their deeds either. But I really hope if they read these lines, they will consider it regarding their next travel.

What are our possibilities of discretion? If you travel and can choose from several destinations, choose the one reachable without flying. If you can, travel by train or boat. If you have to travel by air anyway, choose the closer destination. If you have to fly there anyway, choose the most modern plane. There can be 30% difference between the emissions of an old and a modern plane. If you travel on business, think it over whether you could solve it with online video conference, or whether more negotiations could be held within one trip.

These are all opportunities of discretion with which you can do a lot for Earth while your standard of living does not decrease. We simply have to be aware of the weight of our deeds and make decisions accordingly. Climate change will be efficient if every person takes responsibility for their actions. I am sure to avoid flying in the future and to travel that way only if it is completely necessary. Here's a good example: I recently saw one of Greta Thunberg's videos in which she travelled to a climate conference by boat. I really liked her way of thinking.

11.5.3. Do not smoke (so much)! The example of conscious thinking according to different life cycles

I was driving in Budapest because I was invited by a TV channel where I was asked about environmental topics. I was happy about the request since I really like to take active part in turning the world in a good direction. The driver in front of me threw a cigarette butt out of the car window. I don't want to comment on his action; in my country everybody is free to do that. But this motif gave the inspiration to write this subchapter. **I would like to say that I don't want to start blaming smokers** since I know their habit is also an addiction. I am really aware of the spiritual reasons for addictions and I know very well how hard it is to give up addictions[1]. But even a smoker can choose to smoke less if they can't give it up completely. Each cigarette is a nail in the coffin for their body and worse still for the Earth as well. It sounds much better the other way around: each and every missed cigarette

---

1 Addiction means spiritual dependence. If you are interested in this topic, I truly recommend the In the Realm of Hungry Ghosts by Dr. Gábor Máté or my next book which deals with this topic

is a chance for their body and a breath for the Earth. I know people's instinctive reaction is that it is a too small an action for anything to depend on it. But read on and decide for yourself whether it's true.

The action at the beginning of this section started a train of thought in me. How did the thrown cigarette butt get here? First of all, tobacco was grown, through which a lot of lands were taken from Nature. Then tobacco was harvested, for which vehicles run by fossil fuel were used. These harvesting vehicles had to be produced from a lot of raw materials, using even more energy. The harvested tobacco was shipped (again plenty of emissions) and the transporting vehicles had to be produced as well. Afterwards, tobacco was taken to dryers where a huge amount of energy was used for this process. Then came chopping the tobacco and inserting additives, again using a lot of energy. The next step was packaging. Paper and filters were produced in other factories and were transported to the cigarette manufacturing facility. This was followed by boxing and transporting to the distributor. From there it was taken to the tobacco shops. In the tobacco shop lighting and heating were on, which again generates energy and waste. The customer probably travelled to the tobacco shop to buy cigarettes in a diesel or petrol car. Then, while smoking each cigarette, while the tobacco was burning even more CO_2 went into the atmosphere in addition to a lot of other pollutant material. A cigarette butt should be considered hazardous waste due to its high nicotine, tar and pollutant content, as well as the risk of infection. But we are silent on that topic since it would be difficult to solve it for social reasons, so butts become simple waste. If we walk along the city streets, we can see that a big portion of them become street trash that is blown everywhere by the wind. The problem is that rain washes them into watercourses and pollutants are released there. What's worse is that a part of the butts become micro-plastic as they are slowly worn down by water during transportation. If we grew grains on the lands currently used for tobacco and shared them among poor people, there would be no starvation on Earth. If we gave back the lands used for tobacco growing to Nature, Earth's GHG-binding capabilities would grow significantly. If nobody smoked, Humankind's GHG emissions would shrink. As illustrated by the above life cycle of cigarettes, smoking belongs to the most climate change generating activities of all. If you don't want to give up smoking for yourself, do it for Nature, to reduce the number of smoked cigarettes. If you don't care about Nature, do it for your children's future! Please do it for someone; everybody has to do what they can because we are in really big trouble! **Smoking was a very nice habit, by the way.** When smoking belonged to rituals and celebrations within native American cultures it was a really pleasant habit, but it too was ruined by mass production and profit hunger. People in the developed world want everything immediately and limitlessly, but this way erases the true value that is had when I smoke a pipe or a good cigar on occasions. This example clearly shows the anti-cultural effect of mass production. A nice habit became a massive source of addiction, and it happened because people eagerly took it on. However, it benefits no one except the narrow fraction of the population who are making a profit from it. We need to change this! But the first step is to become aware of the consequences of our actions. The aim of this writing is only that. Then comes the next step, namely that every single cigarette you smoke makes you remember what it means if you smoke. This is followed by reducing your daily number of cigarettes.

Please remember that taking responsibility for our actions is the next step in the fight against climate change. At the same time, smoking is only one example of why it's important to be aware of the consequences of our actions which we can't sweep under the rug any longer. We need to face ourselves in other areas as well.

11.5.4. Three simple habits to reduce your CO_2 emission by more than 10%

The three rules are the following:
   I. Minimize your beef and dairy products consumption.
   II. Minimize buying food which was produced outside your region.
   III. Do not throw food away

These three simple rules result in saving 10 to 15% CO_2 emission in case of an average European citizen. To feel how big a number this is, consider that in 2010 the EU agreed to reduce its CO_2 emission by 20% by 2020 compared to the level in the 1990s. Unfortunately, this was not fulfilled by every state, although there were others who exceeded the requirements. If we keep only these three rules, we can fulfill half or two-thirds of the political commitment without society making any particular effort. It's a really significant thing; we can do a lot with it! Either way, every little reduction is a chance for the world's ecosystem to last a little longer without reaching its final collapse. Any small things we can do, we have to do, without being msled by how small they seem. Eight billion people's small actions taken together result in an incredibly big action.

I have already written about the extremely climate-destructive features of beef production and consumption, so I won't go into more detail. As plenty of cows have to be raised for dairy products, these are also really climate destructive. Beef can be easily replaced by other meats. If we see it from a climate protection point of view, poultry and fish are the most beneficial. The climate gas emission of pig farming is many times higher than that of raising poultry. Tasty and plant-based meat substitutes are also available. I have tasted quite a few of them; the difference is almost undetectable. These meat substitutes have the least impact on climate. We have avoided beef in our family's diet ever since we learned about all this.

Then we started to work on reducing our use of dairy products. This is a bit more difficult to solve. For example, rice milk or coconut milk are not European products, so we would break the second rule if we bought coconut milk instead of cow's milk. Replacing dairy products in terms of taste and consistency is not as easy as replacing beef with chicken. We finally solved ths problem quite well at our family level. We tried a lot of things, and more and more dairy products were removed from our shopping habits. You can buy a plant milk producing machine for example, which is a bigger investment at the beginning but pays back afterwards. We are trying that way nowadays.

Unfortunately, coffee is not grown in Europe, so there are products for which the second rule cannot be kept. However, there are only few products like that. Not long ago I took a package 'Hungarian red paprika' from the shelf in a shop. I read the fine print and it turned out that 90% of the paprika was from Chile and only 10% was from Hungary. The food industry tries to hide traces and sell products in an incredibly disgusting way; we have to pay attention so as not to be cheated.

Unfortunately, to change your shopping habits, you have to spend time and energy to read about the product and where it comes from when you buy it. If it comes from far away, an alternative must be chosen. There is such a wide range of products that we will not suffer from a lack of something if we take care about maintaining the second rule.

When we discard food, it means that it was already produced, packaged, delivered to the shop and taken home. It is loaded by the CO_2 emission of the whole process. If I don't eat it, it goes into the trash unused! It usually breaks down in the landfill under airtight circumstances, which means that mostly methane will be released into the atmosphere and not CO_2. So, if you throw it out, compost it, although it's best if we don't throw food out at all. We should instead give it to the poor, a more climate conscious and a better decision from a human ecological aspect.

To not throw away food requires nothing more than spiritual consciousness. If I am aware of the consequences of my action, it will immediately be a better feeling to eat it instead of throwing it out. In western society, 15 to 20% of food ends up in the garbage. The German president claimed in 2019 that if the food thrown out by the German population in a year was put on a train, that train would be so long that it would reach Russia. In most cases food is discarded because we bought too much and because we don't want it to go bad in the fridge—our biggest problem in our good situation. Every evening I check what is needed to be eaten not to go waste. I prepared scrambled eggs yesterday and put in all of the withering vegetables I found. It turned into a delicious dinner.

There is usually sloppiness, inattention, carelessness or spiritual problems behind throwing out food. If I never want to eat the food which is at home then I am choosing to stay at the level of Desire. But it can also happen that because of my spiritual problems, no matter how hungry I am I don't want anything I already have, and I have no idea what to eat. If you usually feel like that, it is worth digging deep inside yourself, as it's most likely that you are living at a Life-destructive level of consciousness. Since I left behind these levels, I haven't once had the feeling I didn't want to eat anything I had on hand. Before, I had had it many times. If you regularly buy more stuff than what you eat, you may be a shopaholic. Addictions are also the symptoms of the level of Desire.

While a billion people are starving on the Earth, it is pretty absurd that we can't decide what to eat, and even more absurd that we throw out food. Simply put, it's an issue of conscience and ethics to ensure that the food we buy does not end up in the garbage.

11.5.5. Did you ever imagine that composting could be so useful?

The worst thing you can do with green waste is to throw it into the trash! If we compost, decomposition happens in the presence of oxygen so only CO_2 is released into the atmosphere from the decomposing materials instead of the methane released when decomposing in a bin. This makes a huge difference in the warming of Earth's atmosphere. It's worth composting for this single reason, but the good part is still to come.

Throwing organic substances into the bin is a problem since they remain unused—and it is serious wasting. **Did you know that there are as many creatures living in a spoonful of compost as people living in the world?** So, if you compost, you create a living space which is full of really useful, miniature living creatures. These living creatures work constantly on keeping the terrestrial climate in balance. Due to their rewarding activity, carbon is extracted from the atmosphere and gets bound. During humification we actually store carbon in the soil structure. So if we compost and meanwhile make the soil intensively rich in humus, we improve the carbon balance of the atmosphere! CO_2 may be released into the atmosphere in the short term from the decomposition of organic substances (after putting waste into the compost), but in the mid-term you are creating a carbon-binding living space.

One of the main causes of biodiversity reduction is the incredible scale of soil degradation. In just a few decades, the erosion caused by irresponsible human actions and agriculture ruin the soil that had been forming for more millions of years. With composting, you create soil (humus) with which you can work against this process, especially if you put this produced humus into the natural cycle. Composting is a great thing, isn't it?

When I began to compost at home, I explained to my family that we were creating Life with it. Our children happily collect and take out the green waste from kitchen. It's a very good feeling for the family—like regular little good deeds. I made our composter from the wooden planks which were left over from the demolition of an awning. I cut them in a way that they hold each other together, so it can be taken part when it needs to be emptied and afterwards put together again. The idea is not mine: I found it on the internet a few years ago. It's a wonderful feeling that so many valuable resources end up in the end of the garden, and instead of destroying Life we are creating it. I hope you will feel like doing it too!

11.5.6. Five more eating habits to reduce your impact on climate change by 10 to 25%

Feeding about 8 billion people is a really serious factor among the reasons for climate change. It is obvious that we cannot starve in order to save the planet. At the same time, a lot of our consuming habits are an unrealistic luxury from the Earth's aspect. If we care about these things, we can do a lot for our children's future and in most cases we will be healthier. This sub-chapter introduces five highly efficient methods in addition to the ones you have already read about. If you want to take an active part in our common future, choose any of them and you can do a lot without reducing your quality of life:

Habit 1: **Never seafood again.** I do not know how it is with you, but I like seaood and have eaten quite a lot of different types during my life. This turned out to be a mistake, but I am not willing to eat seafood any more. I'm not asking you to do the same, but at least try to minimize it if you can!

We hear it everywhere that seafood is healthy. The good effects of omega-3 fatty acids on health are often mentioned, as is the fact that seafood for the most part constitutes a highly digestible, low cholesterol diet and that it is a tasty protein source. All of these are certainly true. On the other hand, fishing is a 35 billion dollar industry in the USA alone, which spends a tremendous amount of money encouraging people to eat fish. The industry's interest certainly is sharing only the advantages and hiding the disadvantages, but look behind the scenes and see why it is so harmful to eat seafood.

Industrial pelagic fishing has developed to such a high level that thousands of massively productive and effective fishing and processing plants are floating on the seas. These destructive machines destroy the ocean's wildlife faster than ever before. Only 3% of the global tuna population has survived, but the ratio of surviving fish populations in all superior fishes compared to their natural conditions is below 15%. Industrial pelagic fishing is responsible for this, which has several harmful consequences:

- Scientific articles prove that if we continue fishing as this pace, the world's oceans will die out completely by 2048.
- Oceans are responsible for about the 80% of Earth's CO_2 binding capabilities. This massive destruction of the oceans reduces their CO_2 binding capabilities extremely quickly, which will result in the unprecedented acceleration of climate change.
- Boats use nets as big as several football fields put together. These mow down the plants at the bottom of the sea because of their weight, which results not only in the catching of the fish but the destruction of the flora on the sea floor. These boats destroy all the whole marine plants and corals in an

area one-half the size of Europe every year.
- Pelagic fishing is directly responsible for the 50% of the death of coral reefs.
- Slavery and forced labor are still common in pelagic fishing. Murders are also common, as accidents can be easily blamed for the incidents.
- 'Untargeted fishing' means that nets catch whales, sharks and dolphins, which have no meat value. Regardless, they are killed and thrown back dead into the sea because they are competitors. These animals are wrongly blamed for the shortage of fish in the oceans.
- Famine increased in Africa from the shore to a distance of 1600 km inland as industrial fishing vessels appeared in the coastal zone, due to which the local traditional fishermen cannot catch fish. Of course, there are EU fishing companies in the background, which are supported by EU endowments from our taxes, on the grounds of food safety.
- Terrestrial sea fish farming causes incredible environmental damage. The pollution from an average salmon farm is equal to the pollution from a community of 10,000 people. Because of the high concentration of fish, they swim in their own excrement, so the proportion of sick fish is quite high. More than 50% cannot be processed because they die. From the remaining less-sick fish, pâté is made, which we eat heartily.
- Due to bio-accumulation there is a great amount of pollutants accumulated in sea fish. The mercury concentration in sea fish is significantly high, and several pesticide residues are also detectable. Eating a lot of sea fish is not healthy, but rather brings you closer to cancer.
- Fish losses resulted in the reduction of seabird populations by 70%.

- Pelagic fishing is responsible for 50% of microplastics with their discarded broken nets and other equipment. So we will not save the oceans by not using plastic drinking straws, although it is true that many small contributions can collectively make a big difference...

I recommend you watch the movie Seaspiracy about these facts and many other topics in connection with sea fishing, which you can find among the recommendations at the end of this book.

Humankind is not able to survive the destruction of oceans, but if we continue in this way the world's oceans will turn into dead water. Unfortunately, the only way to step out of this nosedive is to stop eating sea fish and salmon in spite of the fact that we love them. If someone is really keen on it and cannot give it up, try to eat sea fish only on special occasions. Of course, this advice is not relevant to poor, starving countries.

Habit 2: **Minimize the consumption of canned food.**

To produce the packaging for canned food takes 10 to 100 times more energy than the food it stores. In case of certain foods like tomato puree, this proportion is much higher. It therefore causes a huge amount of GHG emission to take this product home. If the product is heat-treated, the proportion is worse.

Habit 3: **Give up drinking coffee.**

Drinking coffee is a very popular habit. Unfortunately, in Europe drinking coffee is in the top-ten foods which cause climate change the most. To grow coffee beans a huge amount of rainforest is destroyed, reducing the Earth's GHG binding capabilities. At the same time, the coffee industry can afford that most of the coffee distributed in Europe is delivered by plane—not by boat. The GHG emission of transporting coffee this way damages the environment extremely. Roasting and preparing coffee before drinking it also uses a lot of energy. If we think in terms of life cycles, it's not difficult to imagine why it is so climate destructive to drink coffee. Obviously, the aim here is not to tell someone to give up drinking coffee if they really like it. The aim is not to drink it all the time, but only if we really need it. Not to mention that caffeine addiction can develop fast and is really harmful, so you can do your body good by avoiding too much coffee.

Habit 4: **Avoid eating so much ice cream and frozen foods.**
Ice cream and frozen foods are also famous for their extreme effect on climate change. The extremely high energy need causes the main problem of these products. Producing ice cream can only be done with extremely high energy usage. But think about it: it must be kept cold from the first moment it's produced. In the factory, it is put in a huge cold storage area immediately. From there it is distributed by refrigerated trucks to all over the continent. When it arrives at the stores it is put into fridges. An average ice cream trat or container that you take out from the fridge in the store has been kept frozen for 3 to 6 months. Just imagine how much energy is wasted just for a 100 g delicacy. Locally produced ice cream has a similar enjoyment factor but its impact on climate is only a fraction. We can therefore eliminate it in our life without ruining our Lifestyle. If I were able to do it I would ban the whole industry, but this is not my business.

Habit 5: **Eat as much locally-produced fruit and vegetables as possible.**
We have covered many topics so far about what not to eat, and what to reduce. Now let's talk about what is good to both the Earth and you. Consuming locally-produced and unprocessed fruit and vegetables has the least specific GHG emission, so buy a lot of fruit and vegetables from local markets instead of supermarket chains. Even if you will not be a vegetarian or vegan, the more you eat from these kinds of foods the healthier you will be, the more energy you will have, and the more you will protect the environment. If you compost the vegetable residues, you will do even more. My family goes to the local market every Saturday and buys a lot of fine fruit and vegetables. It;s great for the Earth and also for our bodies.

11.5.7. Change to electricity wherever possible!

Using fossil-based fuels is the strongest driving force of climate change. Anyone who is even slightly interested in climate change knows this. Despite this, people do not really consider how much fossil-based fuel is used in everyday life. If we heat our house with gas, oil or coal, then this is our activity which has the biggest fossil emission. Our stove or cooktop can also have a large emission if it runs on natural gas. Transport is the other activity which uses a lot of fossil fuel. Most people use a gasoline or diesel car, which means burning a huge amount of fossil-based fuel.

So there is a lot of good to be done in our own houses before we start pointing to others. We can of course start with smaller things. For example, I have recently changed my gasoline lawnmower to an electric one. **We should think about internal combustion engines the same way people thought about horse-drawn carriages 50 years ago: they are outdated, obsolete things.** We have to eliminate them from our lives! Everything that works with an internal combustion engine or fossil fuel-based boiler is a climate killer. Besides, there are economical alternatives to all of them.

With the improvements to my house so far (insulation, heating modernization, solar panels, changing doors and windows), I have reduced our CO_2 emission coming from heating and hot water usage by about 70%. But the remaining 30% disturbs me very much, as it is still emission from burning natural gas. The price of solar panel systems has fallen so much that I decided to put an air conditioner in each bedroom and I am going to heat with these in winter and cool in summer. The COP value of these devices is so good that they hardly differ from a heat pump, and you do not have to re-pipe the half of the house again. If I buy enough solar panels, I can heat and cool my house all year long for free. This way I can get rid of natural gas as a fuel forever. We cook with electricity at home. It will be a wonderful feeling when our house becomes carbon neutral in both electricity usage and heating. Our electricity usage is already covered by the solar panels.

It's good to know that as of 2019 solar power generation has become the cheapest method, taking over from nuclear energy production in the 'cheapness' competition.

It is high time to change to e-cars in transport. I have not been able to do this yet due to the driving range problem, but it really disturbs me that I drive 50,000 km a year, burning about 3.5 m^3 of gasoline. To calm my conscience, I compensate for it with tree planting. Despite that, it causes real remorse to me every time I climb behind the wheel. As a company leader I have to drive quite a lot over large distances, and the electric charger network is pretty poor in my country as of this writing (2021). I decided not to change to an e-car until I can make a Pécs-Budapest return trip with minimum charging. This means about 480 km of driving, including getting around within those cities. At the end of 2021, cars with over 600 km range will be launched by more producers, and hopefully this distance can be made with these cars with low-cost driving. Slowly, there will be e-cars available to match my needs. And of course the question of whether it is worth it comes up. I calculated it, and in my opinion it is definitely worth it! The annual fuel cost of my diesel car is HUF 1,450,000, plus the service cost of about HUF 350,000. Insurance and parking fees add another HUF 400,000. So driving is quite a big luxury in my life as it costs about HUF 2,150,000 in an average year. If I buy an e-car and I charge it on my premises with solar panel, I will save 90% of my current fuel usage (sometimes I have to charge it with a fast charger on the way). The service cost of e-cars is half the amount of diesel cars. Parking is often free for electric vehicles. Insurance is cheaper. If we add up all of these, my yearly cost reduces to about HUF 500,000, so my annual saving is HUF 1,650,000. The price difference between a new long-range e-car and a diesel car is about HUF 8 million, plus the price of solar panels is HUF 1 million after the subsidy. The payback period is therefore six years. At the same time, an e-car lasts longer than a diesel car, so we don't have to buy another e-car for 5 to 8 years. If we add to this payback period the conscience benefit and the fact that we do not pollute the air in our immediate environment, it seems to be a fairly good investment to buy an e-car. However, it's useful to think about these figures in terms of your own driving habits.

I could show you a similar favorable calculation in the case of replacing natural gas in heating, but will skip it in the interest of length, but the payback period is longer in my case, which can be reduced by 50% funds by proposal or by an eligible bank loan with 0% interest.

So change everything to electric power, from your lawnmower and boiler to your car, and maximize the number of solar panels you have! With this, you not only do good for Earth and protect your children's future, but you save money!

It is also worth it to mention that the efficiency of an electric engine is above 80%, while an internal combustion engine's is 35%. That means that from the same amount of energy at least two times more kinetic energy is generated. This is the other big advantage of electricity; I can achieve the same mileage with less energy use.

There are of course counter-arguments regarding changing to electric, from which I would like to share some with you:

- *'I will not change to an electric lawnmower because I don't have the patience to carry the cable here and there'.* Yes, well. There are things we have to do in order to have a better world. For example, we have to carry a cable. But I have to add that there are battery electric lawnmowers available at quite reasonable prices nowadays, so it is not necessary to carry a cable any more. I admit that the cable disturbed me as well in the beginning, but I am totally used to it now and it doesn't disturb me at all.
- *'Charging an e-car takes a lot of time'.* This is true! This is the reason why many people can't change. If someone lives in a 4th floor apartment, they can't easily solve the charging issue. But if someone has a business location or lives in a detached house, they can charge their cars fully. In this way, with a proper time and route planning charging at stations can be minimized.
- *'I do not like electric stoves; I can only cook on a gas stove'.* My mother said the same long ago. Then I persuaded her to change to an electric one and she learned to work around the differences in one or two months. She does not miss the gas stove any more.

I personally hated cooking with gas because a disgusting deposit was created on the cabinetry near the stove. I had to clean a lot. This is not the case with the electric stove, and it provides a lot of functions and settings that the gas cooker can't provide.

- *'Heating with gas is comfortable and cheap. I won't change to electric because it breaks the bank'*. This is true only if you do not use solar panels. But if you have enough solar panels, you can heat your home for free. Furthermore, you can cool your house with the same device if it is warm outside. In other words, you can live at a higher standard of living in a climate friendly way—and cheaper! Believe me, it is worth investing in it! Don't forget that there is almost always a 50% state-supported rebate on it. If there is no such offer available to you now, pay attention to the press because these kinds of programs are announced quite often.
- *'It is not worth changing to electricity because of the huge grid losses and power plants'*. There is some truth to this, but it is not an up-to date idea any more. It is true that there is a huge loss from the electricity grid because of the transportation of electricity over long distances and because of the loss from transformers. It is also true that there are a lot of coal and other fossil-based power plants. But in most countries' energy mix the fossil-free energy production is 20 to 80%. Hungary is not in a leading position in this field, but the proportion of green energy production is increasing year by year. At the same time, if you compensate your energy use with solar panels, you are doing everything you can. It is also good to know that from this year on, CO_2 emission will be heavily taxed; that's why it will be worth installing CCU/CCS technologies for fossil-based power plants. That means the CO_2 emitted through the chimney of the power plant is bound immediately before it gets into the atmosphere. This means that in the next decade the whole sector will move in the direction of CO_2 neutrality. It's time to stop gas heating and the ancient solution of the internal combustion engine.

- *'It's not worth changing to an e-car because its environmental impact from its production and disposal is much bigger than a gasoline car's.* This is also partly true, but unfortunately this principle doesn't stand on its own either. Life cycle analyses by several universities found that the distance between 80,000 and 150,000 km is when an e-car makes up for the excess emissions of its production and disposal. Afterwards, every kilometer driven by an e-car is an environmental benefit. As I drive 50,000 km a year, after two or three years my car protects the environment.

I hope I could give you some valuable ideas! By transforming the energy and transportation sectors, we could save 40 to 55% of global CO_2 emission. That is to say, it's well worth cleaning up your life!

11.6. Having and raising children responsibly

You will immediately understand the point of the following sentences based on your readings in the Program of Population:
- Have as many children as to whom you can provide full, selfless adult attention!
- Have children responsibly and consciously! If you are a woman, choose a partner who treats you equally! If you are a man, do everything you can for female empowerment!
- Have a tool for contraception every time!

11.7. Vote responsibly!

Fortunately, most of us live in a democratic country. So if you go to an election, vote for parties:
- who take environmental and climate protection seriously (and not just talk about it)

- for whom democracy is really important
- who do not promote national identity at the expense of other nations or global targets
- who fight for female empowerment and
- who fight for the equality of minor ethnic groups
- who believe in the unity of Humankind and not in the competition between nations
- who consider it important to preserve ethical and cultural values

I know there are no political parties on Earth who take all these things seriously. But you should vote according to your own decisions about parties who are the closest to these values since you can move Earth's future in the best direction by doing this.

11.8. The principle of interdependence on the individual level

Do not forget that interdependence is the basic principle of close relationships, and close relationships are the cornerstones of your personal happiness. Dare to depend on others and dare to be open. Obviously, courage is needed to do this, so it is not by coincidence that this is the first Life-supportive level of consciousness! It works equally well in your private life, family and work..

11.9. The endless possibilities of technical solutions

Being an engineer, I have the most experience in that field, although this book is not about engineering because that the number of technical possibilities is almost endless. Today we are able to live more climate-consciously in so many ways that a separate book could be written about it. You can find plenty of information in this field and you can ask for advice from experts in more and more places, so trust the engineers: they will solve your problems. You just have to set the right goals and take the financial sacrifices that go along with them. Keep in mind that there are newer and newer innovations launched on a daily basis. After information technology, the green industry is the second most rapidly developing industry in the world. There has never been a better time to join the wonderful development that society is experiencing today.

11.10. Protect, take care of, respect and support Nature

This whole book is about it! Now you clearly know what the title means. I've highlighted it here again just for the sake of completeness. It cannot be emphasized enough..

11.11. The power of individual and group meditation

Meditate on a daily basis for the strengthening of the remaining wildlife on Earth. If you can, do it in a group meditation!

11.12. More than a hundred practical possibilities

In this section I am listing more than a hundred possibilities of what you can do to make your life better and take an active part in having a better future for Humankind. Imagine that if each of these actions reduced your GHG emission by only 1%, what would happen if you implemented all of them in your life. My recommendation however is to be gradual. It's difficult to change our habits and in order to something to be second nature, we have to train our brain for months. And it needs attention. I always add one to five things to my New Year's resolutions that I plan to build into my life for the sake of climate protection, and I try to ask my family to do the same. This way, we live more and more carbon neutral and happier. I started doing this more than ten years ago so I know it for sure that it works. Every little step counts and every change that seems small leads Humankind's development in a better direction.

And now here comes the more than 100-entry list, broken down into topics:

Transportation:
- Use more bicycle, e-vehicle or public transport in the city.
- Go on foot whenever you can.
- Climb the stairs; use the elevator rarely.
- Always favor trains and ships on longer distances over using buses or planes.
- Whenever you can, use telepresence.
- If you plan to go on a holiday, travel to the nearest destination among your holiday ideas.
- Use planes only if absolutely necessary.

Buildings, energetics, GHG emission reduction
- Do not use air conditioning, or minimize it, unless it works fully with renewable energy.
- Do not throw out fridges or air conditioners without proper disposal of the gas inside! These devices contain GHG thousands of times stronger than CO_2!
- Insulate your home.
- Change your doors and windows to highly insulated ones.
- Use energy-saving devices.
- Use LED lighting.

- Use and produce as much green energy as you can.
- Do not heat or cook with natural gas, coal or other non-renewable resources.
- Do not heat or cook with wood.
- Do what you can to minimize your and your family members' pollutant emission.
- Adjust whatever you can in your life to weather conditions!
- Install a white or really bright roof.

Shopping habits
- Buy the products produced nearer to you.
- Buy more durable products.
- If something breaks, have it fixed.
- If you do not need something that is still working, donate it to people in need.
- Prefer buying second-hand products.
- Don't buy anything you don't need!
- Produce by yourself everything you can.
- Rent what you can.
- Reduce or heal your shopping addiction.
- Do not buy the products of companies engaging in anti-Human or anti-Nature activities if there is an alternative product. Warn others as well.
- Do not buy products that are advertised ignoring the ethical rules.
- Choose white or really bright versions of cars and other products you use outside.

Respect and protection of Nature
- Plant and divide. Create habitats.
- Protect Nature and encourage others to do so.
- Plant at least 20 trees every year! Take care of them afterwards. If you can't do this, hire a professional organization.
- Spend as much time in Nature as you can.

- Buy lands and hand them over to Nature.
- If you have unnecessary lands, give them to Nature.
- Take part in community garbage collection, tree planting and other activities.
- If you see that somewhere Nature is being destroyed or harmed illegally, report it to the authorities. Do not be indifferent.
- Compost and return the valuable compost to Nature.
- Do not use artificial flowers.
- Decorate your home and office with natural plants.
- Support environmental organizations.
- Accept that every living creature has the same strong right to Life as you. Live according to that!

Spirit and climate protection
- Do not let longing rule your life.
- Mitigate your hedonism consciously.
- Mitigate your selfishness and vanity consciously.
- Do what you can to heal your spiritual problems and addictions.
- Help others on their way to spiritual development.
- Deepen your self-awareness.
- Make spiritual development the main target of your life.
- Make a balance sheet of your Life-supportive and Life-destructive actions.
- Enhance your Life-supportive actions and reduce your Life-destructive ones!
- Experience the joy of selfless donating. Do it regularly as part of your life.
- Heal your inner fears; and don't flee from them.
- Take responsibility for the impact your actions have on climate. Be aware of the consequences of your actions, and set an example with them.
- Be happier, more balanced and more peaceful.
- Dare to trust in others; dare to think in communities.

- Trust in the principle of interdependence.
- Be rational and material only in cases when it is really necessary.
- Listen to your feelings and intuition.
- Develop your creativity.
- Dare to be yourself; be honest with yourself.
- Face your inner problems bravely.

In society as an individual
- Support every activity which is based on the cooperation of several communities.
- The 'act locally, think globally' principle should prevail in every of your actions.
- Speak up against the current unequal economic system.
- Do what you can for democracy.
- If you live in a country where there is population decline, don't let the related propaganda scare you.
- Don't let the fake happiness images proffered by society mislead you (it's not possessions and externals that count).
- Respect science.
- Respect older people's wisdom.
- Live according to Human-ecological norms.
- Live according to environmental ethics expectations.
- Respect religions, including ones other than yours.
- Do not possess.
- Speak up for peace, climate protection and acceptance of diversity.
- Fight for female empowerment.
- Fight for any minority's equal rights.
- Do not accept the demagogic side of national identity.
- If you are an inventor, make your patents public property.
- Believe in Humankind's unity.

Having children and raising children
- Care about contraception.
- Plan the number of your children consciously.
- Have only as many children as to whom you can provide enough selfless attention.
- If you are a man, support the equality in your relationship.
- Involve grandparents and great-grandparents in parenting and family programs.
- Be with your children as much as you can in such a way that you pay attention to them absolutely selflessly.

Media, culture
- Avoid advertisements and other marketing tools that cause desire or longing.
- If you learn about the anti-Human or anti-Nature activities of a company, share that information with others on social platforms.
- Increase the Life-supportive media content in your Life and reduce the Life-destructive content.

Water management
- Live in a water-saving way; use water-saving systems.
- Make use of rain water.
- Make use of grey wastewater.
- Take care that your wastewater gets into the environment only after proper treatment.

Eating habits
- Eliminate or reduce beef and mutton in your diet.
- Consume fewer dairy products, or if you can, give them up.
- Do not buy food coming from other continents, or minimize it.
- Eat as many local and unprocessed fruits and vegetables as you can.
- Minimize the eating of canned and non-perishable foods.

- Minimize the eating of ice cream and frozen foods.
- Minimize coffee drinking.
- Do not buy foods containing palm oil.
- Reduce eating sea fish.
- Do not throw out food. (Don't buy unnecessary food!)
- Favor the food produced the closest to you.
- Do not smoke, or at least cut down.
- Do not drink alcohol, or at least cut down.

Waste
- Reduce the amount of waste you generate.
- Think of the whole life cycle of a product.
- Collect waste separately; help recycling efforts.
- Buy low-waste products.
- Never burn waste.

If you've read these 100-plus possible activities, you probably found a lot which are already parts of your life. Be proud of them and set an example. Think about whether those activities you have not yet implemented in your life yet, and how many of them cost nothing or need just a little financial sacrifice. It is not true that environmental protection is a 'sport' for the rich; **anyone can act against climate change according to their possibilities!**

CHAPTER 12

The implementing steps of the 6 Programs in the short, medium and long terms

My vision, as introduced at the beginning of the book, is a transition to a community in which People live in balance, peace and harmony with themselves, each other and Nature. There isn't even any need for money in this future world. The social system is based on creativity and motivation. There is no deprivation. It is every Human's individual right to satisfy their physical, mental and spiritual needs which are facilitated by science-based technological development. As there is no deprivation, there is no fear or possession. People consider mental, scientific and spiritual development to be their main mission in life.

We are already prepared for this future world scientifically and technically. Old habits and wrong thinking are the only reasons why we don't already live as described above. According to my estimation, we could reach this wonderful ideal in about 3 to 5 generations. **I trust you can believe in it as well.**

Obviously, we will never live in a perfect world, but we really can approach what I've described. How? This book is all about it, and the further books in this series will be about it too. In this chapter I am introducing—broken down into steps—how the 6 Programs are built up and interact with each other, and how they will take us there in 3 to 5 generations. Our generation's task is to start on this path as soon as possible since we are running out of time. We are approaching the line of destruction from which there is no turning back. **Our generation's task is to buy some time for the next generations. It's time for everybody to do everything they can according to their opportunities!** I have written down the implementation steps in chronological order. I am describing the process from the perspective of a newly set up agglomeration because if we wait for the relevant global agreements to be in place, it will definitely be too late. **This whole system can only start as a bottom-up initiative** since community leaders will realize they can't provide the future for their citizens any other way. Once there are many agglomerations representing enough power, the process can be sped up by setting global directives.

Now let's see the fastest scheme for the first agglomeration to be set up, whose example will be followed by many others:

In the present
The first agglomeration in the world gets set up from the fusion of some existing settlements. They establish their local government by delegating the local governments of the member settlements, and the decision-making work gets started.

Steps after establishment:
- Draw up in detail the educational directive based on the basic principle of population control (Program of Population) which must be expanded to a similar restructuring of the education and child care systems (Program of Happiness, Program of Society).
- Set up the Natural Area Cadasters (Program of Revitalization)
- Draw up the Agglomeration Directive and in parallel the Natural Territorial Development Directive (Program of Agglomeration). To be able to count the agglomeration's emission balance, the calculation method of assimilation capabilities must be worked out.
- Draw up the Agglomeration Ethic Codex.
- Accept both directives and codex (Program of Society).
- Work out the Service List with its schedule, economic and social incentive systems (Program of Society).
- Work out the general mental hygiene system and the kinesiological training of general mental hygiene specialists (Program of Society)

- Work out and accept the counting method of GHI, its measuring and data managing system (Program of Happiness)
- Cancel the principle of infinite growth from the economic models and work out new economic directives and methods based on the principle of perfect balance. These must be accepted and implemented in economics education (Program of Economy)
- Develop the technical conception of the Global Grid, completed with a global renewable energy development program (Program of Society). Send this as a proposal to all countries and introduce it at the next climate conference.
- Draw up and accept the defense directive of areas defined by the Natural Area Cadasters (Program of Revitalization)
- Develop and accept the Media Ethic Codex about avoiding advertisements and other media content that enhance or support selfishness.
- Develop and continuously organize programs aiming toward community movements, communal life, selflessness, strengthening social togetherness (Program of Happiness)
- Introduce meditation, psychology, self-awareness and environmental subjects in the educational system (Program of Happiness)
- Make the list of Life-supportive human activities and always add new ones. Promote and propagate it. Make a list of Life-destructive human activities, update it continuously and inform people about it.
- Work out a method that counts the balance of Life support and Life destruction individually; educate and train people to strive to manage their lives in the direction of Life support. An advisory and social supportive system must be established for that purpose (Program of Happiness).
- Support the individual search for happiness by providing advice and psychological assistance through the general mental hygiene system.

- Intensely deal with socially motivating and promoting programs focusing on People's spiritual development (Program of Happiness).
- Introduce the principle of a Sun society, promoting and propagating it socially. Introduce taxes and other economic incentives which lead development in that direction on both the corporate and individual levels (Program of Society).
- Begin teaching psychosomatics at a higher level at medical universities and instill it deeply into the public consciousness with promotion methods. The institutional system of healing psychosomatic diseases must be strengthened with it (Program of Society).
- Work out the labelling regulations of products and services.
- Work out and announce the implementation and scheduling of taxes introduced in the Program of Economy.

10 to 20 years later:
- Work out the institutional system of the Population Control Directive, implementing it in the education system and propagating it widely.
- Promote the proposals of the directive with a thematic system of training and community programs.
- Continuously transition the education, child care and family care systems in such a way as to let the basic principle of population control prevail (Program of Population).
- Introduce the Service List according to accepted scheduling through setting up an institutional system and expanding it continuously. The list of activities coming into force will also be expanded continuously (Program of Economy).
- Calculate the balance population number and real population number every year; add its feedback to the strategic plans and regulations dealing with population number. Global values and the correction of global targets result from the summary of agglomeration data (Program of Population).

- Register the GHI in every agglomeration; later the global GHI can be counted from it and feedback to develop strategic plans and regulations (Program of Happiness).
- Introduce labelling.
- Levy the taxes of the Program of Economy.

About 24 years later:
- Give transnational rights to agglomeration institutional systems at this stage only in agglomeration environmental and climate protection issues (Program of Agglomeration)
- Establish the institutions needed for the protection of lands by Natural Area Cadasters and for the development of other natural areas, together with financing.
- Support and strengthen community initiatives politically and socially.
- The Cadasters must deal with the connection plan of natural areas in the long run (Program of Revitalization). Nature must be given rights, and the newly established institution must be capable of providing the legal protection of Nature. The Directive of the Legal Protection System of Nature must be drawn up (Program of Society).
- A detailed review must be conducted by every agglomeration regarding the current condition which determines to what extent the current emission level differs from the assimilation capabilities of the surrounding areas (Program of Agglomeration).
- Every agglomeration must calculate its GHI and work out a long-term plan on how to boost People's happiness (Program of Happiness).
- Determine and relate the balance population number to the population number in every agglomeration. A strategic plan must be drawn up which rules and supports the evolution of population in every agglomeration (Program of Population).

- Establish the Global Parliament, Global Council of Wisdom, Global Scientific Panel, Global Peacekeeping Force (when there are enough agglomerations for this initiative). These organizations do the conceptual governance of the agglomerations' work. The Global Peacekeeping Force takes an active part in maintaining the protection of natural areas (Program of Society).
- Accept the conception of the Global Grid globally at a climate conference. Work out the schedule of its establishment with international cooperation (Program of Society).
- Every agglomeration has to work out a mid-term strategy which includes the plan on how the agglomeration will reach balance with Nature by 2050 (Program of Agglomeration).

30 to 50 years later:
- Realize the protection of natural areas defined by the Natural Area Cadasters and their continuous maintenance, with a special focus on encouraging and supporting the public initiatives at the institutional level. Start the revitalization and expansion programs of natural areas in accordance with the agglomeration delimitation (Program of Revitalization).
- Social promotion of the love of Nature in education (Program of Happiness).
- Establish the Global Grid and the institutions necessary for its operation (Program of Society).
- Expand the rights of every agglomeration beyond the environmental tasks to the tasks in relation with global society and issues concerned with the global happiness program. Set up an independent police force and army in every agglomeration. Transfer national rights continuously to the agglomerations (Program of Society).

- Set up the Global Peacekeeping Force through delegating agglomerations (Program of Society). Set up a global agglomeration control system with the aim to summarize data on the agglomeration level globally, draw conclusions and give feedback to agglomerations (Program of Agglomeration).
- Draw up the mission system of agglomerations and maintain it continuously (Program of Happiness).
- Every agglomeration carries out the developments and changes of their mid-term strategies (Program of Agglomeration).
- Operate and develop an agglomeration analysis system (Program of Agglomeration).

Last tasks for the final transition:
- Final transition to a service-based society. The progressive transition from companies to trade unions.
- The elimination of money. Transition to motivation-based society.
- Society works according to the principle of the Sun society.
- The constant growth of the number of agglomerations in balance with Nature. Reach balance with Nature globally.
- Steady increase of GHI.
- Choose the world language and accept it internationally (Program of Society).
- World peace.

Closing remarks

Dear Reader: I truly thank you for reading this book! If you are reading these lines, it means you have honored me with your attention. I hope I was able to evoke Life supportive thoughts in you that will generate changes in your life. If so, then you have made writing it worthwhile! But it is also important to mention that constructive criticism and creative thoughts are the best friends of creation. So if you have any such ideas in connection with this book, please let me know and contact me at the e-mail address at the beginning of the book. This book can be expanded and become more refined with your help, thereby helping people more and more efficiently. I would also like to ask you to share your opinions with others and do what you can for the development and spreading of the system!

Many things I have dealt with in this book are not worked out in depth. The reason is that I wanted to show a framework simply and briefly for clarity's sake. There will be books among the future volumes in this series which will deal in depth with the topics mentioned. I hope you will join me then as well.

The good thing about this system is that it is open-source, so **anyone can shape, form and develop it as they like.** I really hope that a reflection wave will start which will develop this system through people's endless creativity. I am going to continue my work in case there may be an expanded and developed copy in the future.

Acknowledgements

We have an effect on people during our lives, and many people affect us. The result of all of these effects is what we are right now. By even those who have done us harm have helped us become more and better, so as a matter of fact we could be grateful for the wrong they did to us. That's why in this section I want to express my gratitude to all the people I have ever contacted directly or indirectly. Of course, there are a lot of people I would like to point out here, but because of constraints on the length I have to narrow down the long list of gratitude which is in my mind and spirit. I apologize to those who are not mentioned here!

Some authors and lecturers have had an incredible impact on me; and even though they don't know me personally, I am highly grateful to them. I would like to highlight some of them as a sign of my gratitude and honor: Dr. Joe Dispenza, Eckhart Tolle, Dr. David R. Hawkins, Eric Berne, Jacques Fresco, James Redfield, Neale Donald Walsch, and Paul Hawken.

Besides the support of my family, loved ones and friends, I would like to highlight the three people who had the biggest impact on my life: Dr. Zoltán Egerszegi, Ildikó Tönkő and Dr. Ferenc Szilágyi. I hope the energy of my gratitude will help them many times over, and that I can repay the abundance of selfless good I received from them.

I would like to thank everybody who contributed to the development and publishing of this book either consciously or unconsciously, actively or passively.

But most of all, I am grateful to all those people who take and have taken an active role in Humankind's biggest challenge: the fight against climate change!

Literature

BOOKS

- **Angus Forbes:** *Bolygónk globális hatósága- Hogyan védhetjük meg a bioszférát.* Pallas Atthéné Kiadó, Budapest, 2019.
- **Beau Lotto:** *Deviate.* Libri Kiadó, New York, Hachette Books 2017.
- **Christopher Stone:** *Should trees have standing? Law, morality and the Environment* – source: Molnár László: *Legyenek-e a fáknak jogaik?* Környezet etikai szöveggyűjtemény – Typotex 1999, original source: Southern California Review, 1972.
- **C.G. Jung:** *The Archetypes and the Collective Unconscious.* Princeton University Press, New Jersey, 1990.
- **Dr. David R. Hawkins:** *Power vs Force.* Hay House UK Ltd. London, 2014.
- **Dr. Máté Gábor:** *In the Realm of Hungry Ghosts.* North Atlantic Books, 2010.
- **Dr. Joe Dispenza:** *Becoming Supernatural.* Bioenergetic Kiadó, Hay House Ltd., London, 2019.
- **Eckhart Tolle:** *A New Earth.* Penguin Books. London, 2016.
- **Eric Berne:** *Games People Play.* Grove Press, 1964.
- **Pope Francis:** *Laudato Si' On care for our Common Home.* Encyclical, 2015.
- **Jacques Fresco:** *Designing the Future.* The Venus Project Inc.
- 2007.
- **James Redfield:** *The Celestine Prophecy.* Transworld Publishing Ltd. UK, 1994.
- **John Bradshaw:** *Healing the Shame that Binds You.*

Health Communications, 2005.
- **Klaus Werner:** *Schwarzbuch Markenfirmen. Die Welt im Griff der Konzerne*, Deuticke Neuauflage, 2014.
- **Mike Berners-Lee:** *There is No Planet B – A Handbook for the Make or Break Years*, Cambridge UP 2019.
- **Muriel James – Dorothy Jongeward:** *Born to Win*. Da Capo Lifelong, 1996.
- **Neale Donald Walsch:** *The Complete Conversations with God*. Putnam, 2005.
- **Orvos-Tóth Noémi:** *Örökölt sors – Családi sebek és gyógyulási útjai*. Kulcsjuk Kiadó, 2018.
- **Paul Hawken:** *Drawdown*. Penguin Book, 2017.
- **Paul Ekman:** *Emotions Revealed*. Orion Publishing Co, Budapest, 2004.
- **Zsolnai László:** *Ökológia, gazdaság, etika* Helikon kiadó, Budapest, 2001.

INTERNET SOURCES

- **Globális felmelegedés:** https://hu.wikipedia.org/wiki/Glob%C3%A1lis_felmeleged%C3%A9s#cite_note-Prentice_et_al._2001-16
- **IPCC:** Special Report on Emissions Scenarios - A Special Report of Working Group III of the Intergovernmental Panel on Climate Change. 2000. ISBN: 9780521804936
- **Timothy M. Lenton, Johan Rockström, Owen Gaffney, Stefan Rahmstorf, Katherine Richardson, Will Steffen & Hans Joachim Schellnhuber:** Climate tipping points — too risky to bet against. Nature 575, 592-595 (2019) doi: https://doi.org/10.1038/d41586-019-03595-0
- **Marinov Iván:** Einstein és a méhek. 2013. https://www.urbanle-gends.hu/2013/06/einstein-es-a-mehek/
- **Európa Parlament:** Veszélyben a méhek - Környezetvédelem- 18-11-2008 - 12:13 https://hu.euronews.com/2019/05/20/veszesen-fogynak-a-mehek
- **Tudatos Vásárló:** A magyarok az EU átlagnál is jobban tarta- nak a klímaváltozástól. 2019.09.17. https://tudatosvasarlo.hu/magyarok-eu-atlag-tartanak-klimavaltozas-felelem/
- **Emrod:** Emrod vs. Tesla's Long-Range Wireless Power Technology., 2021 Február 21. https://emrod.energy/emrod-vs-teslas-long-range-wireless-power-technology/

Movies about climate change which I highly recommend

- David Attenborough (2021): Breaking Boundaries: The Science of Our Planet

- Seaspiracy (2021)

- Kiss the Ground (2020)

- Poisoned Soil (2020)

- David Attenborough: A Life on Our Planet (2020)

- Al Gore: : An Inconvenient Truth I. (2006)

- Al Gore: An Inconvenient Truth II. (2017)

- Leonardo DiCaprio: Before the Flood (2016)

- Leonardo DiCaprio: Ice on Fire (2019)

- David Attenborough: Climate Change – The Facts (2019)

- Yann Arthus-Bertrand: Home (2009)

- Josh Fox: How to let go of the world (2016)

Appendices

Every reader has different opinion about spiritualism. There are people deeply interested in the topic but others are not ready to leave the ground of rationality, so they reject this issue. These appendices are for readers who are more deeply interested in the main notions used in this book in connection with spiritualism than what's found in the main text. The following pages are therefore about ego, addictions and the levels of consciousness in more detail. **A separate volume is in progress** which is going to provide new information about the ways in which we can raise our level of consciousness. The aim of that book is to enable bringing together our search for happiness and the fight against climate change in our life. We should start the Global Happiness Program together with it. I hope you will honor that volume with you attention too..

APPENDIX 1

More about ego

The notion of the ego has been raised many times in this book. As I don't know what kind of knowledge readers have about this topic and who might be interested in this basic issue of spiritualism, I have written my thoughts and experiences here in the appendix. I hope you will read these valuable lines and that they will have a positive impact on your life. If you're interested in the topic, you can find a lot of exciting writings in my blog. The address is at the beginning of the book.

Observing the (your) ego

It's always easier to notice others' unfavorable spiritual processes than our own. Why is this so? The answer is really easy, but if you haven't dealt with it before, it may seem strange at first. **Your ego is the parasite of your spirit.** I read this expression from Eckhart Tolle, whose writings I highly recommend to everyone who wants to develop spiritually. Most people reading these things get fed up or have an aversion to them—and that's what proves they're true! Please read on to see the big picture and bring some ease to your life.

We should begin with the meaning of the bolded sentence above, and then I will deal with the evidence. A parasite is a living creature which uses the host in a way that in most cases it does not even realize. Think of the mistletoe that grows into an oak tree. The oak thinks it's part of its body and nurtures the mistletoe. The mistletoe causes no benefit to the oak; indeed, it accelerates its death and increases its suffering. The oak is not aware of the fact that mistletoe is not itself, so it nurtures it until the oak dies. If it were aware of the fact, it would close out the mistletoe from itself! Your ego is just the same. It's not a plant, but a negative energy pack. It embeds itself into your spirit and makes you believe that your ego is Yourself! Afterwards it doesn't care about anything else but makes you believe how important its existence is, and doesn't care about what's really good for you. As a parasite embedded in your spirit, it's engaged only with its own existence and importance. And here is the problem: the existence of your ego is not your existence since **your ego is not You!** When I first read about this, I asked myself: if my ego is not me, then who am I? I was pretty skeptical at that time. My ego made me believe so seriously that my ego is me, and that I didn't have the faintest idea who I was. Ask yourself the same question, and think it over. I needed months to get the answer.

The ego is a negative energy pack existing on a spiritual level that is responsible for your spirit's unhappiness. Meanwhile, it makes you believe just the opposite. It also makes you believe that your ego protects you, it's your ego you can thank for your existence, and it's also the ego you can be grateful to for putting where you are in your life. Take care: these are fake notions! You receive these notions only because your ego is concerned about its existence and tries to strengthen these thoughts within you.

A deeper level of self-awareness comes when you realize that your ego is not you. Why is this so important? Because if your ego manages all of your spiritual processes, this is why you feel unhappy! In other words, the truth is just the opposite of what your ego suggests to you. As your ego cares only about its existence, it has to suggest (as you are the host) that you will not exist without your ego, or that you would be incapable of life without it.

Now let's go back to the beginning: why is it easier to recognize other people's harmful spiritual processes than our own? Because your ego hides everything from you that can be negative about itself! At the same time, the ego is critical of others because if you find negative features in others, your ego uses this to strengthen its existence. You always have the impetus to compare yourself to others, don't you? This is your ego, which always compares.

To attain self-awareness, you need the realization of how strong your ego is and how deep it is embedded in you. Since I have been observing people from this perspective, I can see that almost everybody has this issue. It can also be said that the more hurt one's spirit is, the stronger an ego they have. Don't be misled by the fact that a lot of economically successful people have strong egos. This is because their self-assertion skills are very strong, but inside—in most cases deeply suppressed, without their being aware of it—there are very deep pains in their spirits. These people are unhappy, or their happiness is a self-deception envisioned by their ego. They compensate in front of the world and themselves with their strong egos.

I also had an extremely strong ego. The stronger it was in me in my life before, the more I believed I was on the right track, but the unhappier I became. Now I see crystal-clearly how different my ego and I are, and I am working on the reduction of my ego. I am happier as a result. Unfortunately, there are situations even today when my ego switches on and does something I regret later on. But this is not a problem; it is a good source of self-awareness when we're aware of it. This way, we gain the chance that our ego won't switch on again in a similar situation. It's practical to get rid of our ego slowly. In my case, this process started three years ago, and I still have a long way in front of me. However, it's also true that when Eckhart Tolle (who is said to be one of the most enlightened people in the world) realized that his ego was the center of his misery, he simply broke up with it. He was so disgusted by his ego that he managed to do it. I don't know how he did it, as I can only do it slowly, step by step. One thing is for sure: both he and I know that this is the right way. The weaker your ego will be, the happier your spirit will be, since a weaker parasite will be abusing it.

(Your) Ego's features: separation

I am dealing with a characteristic feature of the ego in this sub-chapter. To get started, I'll write a sentence for you and I'll ask you to please observe your feelings while reading it. Observe only your feelings and reactions:

'I feel that I am perfectly equal to every person.'

What kind of feelings did you have while reading this sentence? Perhaps you felt that this statement was perfectly true? Or rather that it is not true? When you felt it wasn't true, did you feel in how many things you're better, more, or different than others? Or how you felt weaker, and how much less you are than others?

Those who can absolutely agree with this sentence have a very weak ego or none at all. I congratulate you! There are very few such people in the world. If you belong to them, you are a happy, peaceful, balanced person!

But most people will have other reactions to that sentence. Most people's feelings focus on the differences and feel aversion when reading this sentence. This feeling of aversion is the evidence of the ego in your spirit. I too have felt this way (unfortunately).

Unfortunately, the ego is the root of almost all the bad things that happen to us, although it makes us believe just the opposite. The ego says it protects you from every bad thing that can happen to you, but it's not like that at all! The ego only keeps you in fear to make you believe you need it.

Our life has two main periods. Until we are middle-aged we search for our place and want to leave a mark on the world. Only a few people realize the many harmful effects of the ego during this period. The strong desire for self-assertiveness stabilizes and strengthens the ego and the person goes along with it. However, in the second period of our lives we realize that we had been on the wrong path. Spiritualism and spiritual development become the main thread of our lives and we learn how silly and unhappy it is to live our life hidden behind the ego.

The ego enhances the differences between people in our spirit. The ego always compares us to others. It always searches for the ways in which we are better or different. The difference can be positive and negative as well. For example, if someone is more attractive than me, my ego switches on my jealousy and makes me yearn. It can also happen that the ego tries to explain that the attractive appearance of the other person is too much and enhances my difference from them. The ego can also make me feel subordinate to others, which is the fixation with negative difference. As I was really lacking in confidence, I tended to look at others as gods while I considered myself worthless. At the same time, when I was successful in something I was likely to think too much of myself compared to others. Due to my lack of confidence, these were short periods, always followed by some disappointment. My ego appeared at that point and urged me to protect myself next time, and thus strengthened itself within me. The ego does the same with everybody; it uses every emotional injury to prove the importance of its existence and to get permission to strengthen itself within their spirit.

There are people who scramble for money, power, fame or an extraordinary appearance because of this, and the list is long. The ego feels safe due to these, but the strengthening ego asks a high price in exchange. We believe that we're different from others, and meanwhile our relationships become more shallow, so we become more lonely inside. If we aren't honest with ourselves, why would our relationships be? Snobbery and conceit, for example, are strong manifestations of a strong, callous ego. In that situation we have already separated from our real selves and have no knowledge about who we really are. This is the level where our ego has already separated us from our real personality completely and we already believe that our real selves are the ones adjusted to some external demand system. Every snobbish person thinks that they are unique, while most of them are typical and stereotyped. Of course, they can't admit this until they can look behind their ego.

So the stronger our egos are, the less we know who we really are, and the less honest we are with ourselves. The price you have to pay is spiritual loneliness. How many people are lonely inside while living in a relationship? How many people exist in workplaces or school communities while being lonely inside? How many people go out and hang around with friends while being lonely inside? This inner, tortuous solitude is the price of the strengthened ego. This is the spiritual separation from our real selves and from others. This is spiritual distancing. This is the reason there is no real happiness with a strong ego. Our shallow relationships change fast. Only new things can mean an escape from solitude.

Our ego, of course, often makes us believe we are happy. When I buy a better car, bigger house, attain more power or travel farther, the ego says that this is happiness. We feel joy temporarily, but then comes a new desire because something is left empty inside. We escape from the suppressed solitude to a new possession or fulfilling some new desire which strengthens the ego. The bad news is that this process will have a terrible end.

Politicians with infinite desire for power have ended up committing suicide, in prison, in a mental hospital or being executed when their power suddenly collapsed. This is a typical manifestation of the collapse of an incredibly callous ego. Think about the execution of Nicolae Ceaușescu in 1989; if the ego is too callous, you can get rid of it only with giant failures since it will break unexpectedly through incredible suffering. My ego has collapsed three times. Unfortunately, after the first two cases I raised a stronger, more capable and cautious ego. My ego made me believe that everything would be better from then on. After the third collapse, I didn't want to fix it again. That time, I clearly saw the malfunctioning of the ego and its spiritual parasitic system.

Why the ego-free person is more efficient, healthier and happier than the one who has a strong ego, or the 9 tricks of (your) ego

I used to live in the world of ego. My ego was strong and presumptuous. I was convinced that everything was the best, most correct and most efficient only in the way I thought and wanted to carry it out. I was incredibly stubborn and I always knew the right way. I was also convinced that I didn't need external help from whatever you call it: God, the Almighty, the Universe, luck etc. I thought these were only for weak people. I believed I was so strong that I could overcome the hardships of life completely on my own.

My whole life was a real struggle with a lot of anguish and deep falls. When I was down in the dumps spiritually, I suffered a lot. To live in the ego's world is like watching the world through a tube; the hole at the end of the tube narrows down your vision. Due to the ego, we become incredibly purposeful and focused about a small detail of reality. We are convinced that what we experience is the only existing thing, and that anyone who feels different or experiences different things is not normal. We believe the only truth is what we experience. We close ourselves in a narrow-minded and presumptuous world due to our ego, which destroys not only our lives but reduces our efficiency. How can it be that a person with a strong ego is convinced of exactly the opposite?

Dr. Joe Dispenza was once asked during a lecture how he can be so shy when he became such a famous and big personality? He is actually one of the most famous spiritual leaders in the world. He answered that "it was an incredibly huge job to destroy my ego. I really don't feel like doing it again and again!" The ego always wants to grow back in our spirit. It is a spiritual parasite that needs a host. The ego can't exist without you, but you can exist without the ego. Just imagine a living entity in which parasites live. Can it be more efficient, healthier, happier than one that is free of parasites? Which tree grows to be more lush and vital in your opinion: the one infected with mistletoe or the healthy one next to it?

But how does the ego-parasite make us incapable while convincing us just the opposite? I would like to answer this by sharing a few more aspects:

1. Tunnel vision

I have already mentioned it, but let's fix it: if I see only a small part of the world and I am convinced that this is the only existing truth, I will miss a lot of opportunities without even realizing it. If I can choose from fewer opportunities, my efficiency reduces dramatically as it can happen that I reach my goal in the most difficult way while I don't have the faintest idea about the possible shortcuts. I didn't understand why I had to work ten times more than others to reach the same goal while others would just get it. It felt unfair, but it was only because of my callous, hard ego that it happened to me all the time. I felt unlucky and miserable, which made my ego grow stronger.

2. The energetic effects of useless thoughts

The ego keeps you in constant fear since this is the way in which it can convince you that the ego is the only thing that protects you. This is why your brain is constantly monitoring the possible alternatives of your future and you try to choose the best direction. At the same time, 99.99% of your thoughts from analyzing the future will never happen! So 99.99% of your thoughts about your future are absolutely useless. This means that only 1 out of about 10,000 thoughts makes sense. Those without ego trust in the future, so they don't scan it all the time. They can use these 9999 thoughts for things which are much more meaningful. Brainwork burns more energy than running. It's not surprising that after some creative work which requires a lot of thinking we are so hungry. The ego also adds the push to think about the past. If you do something wrong it makes you think it over a thousand times so as not to do it again in the future. It again strengthens the ego in you, but the point is that 99.99% of your thoughts about the past are absolutely useless. This is an incredibly large amount of time and energy.

I used to struggle with a lack of energy when I had a strong ego, but I didn't understand why. It's obvious today, though. I wish someone had explained it to me like a form of help. Well, I'm trying to help You right now! I felt I didn't have enough energy to reach my goals, but I forced myself and proceeded on to my goals beyond my ability. This of course took me in a downward energy spiral. Waking up was a struggle every day. Working more and more every night was torture, but I did it. The myriad useless thoughts drained my energy. Now I can accomplish more things in 8 hours than in 16 before, not to mention that I am hardly ever tired. I am full of energy when I wake up in the morning. I feel good almost the whole day. I can always concentrate on the task in front of me.

I used to plan my days strictly. If someone or something wanted to distract me from my original schedule, I was angry and furious and kept on doing my things according to my original schedule. The anger, fury and forced schedule used up a lot of energy. And why? Because I was convinced that the future would be right only as I imagined it. What a big dope I was ... Life takes care of everybody who is open to it!

It is an energy system that penetrates everything and manages every aspect of life. Only the ego is blind to it, but You or anybody can be open to it! It is ego and exaggerated rationality that push you out of it. It's not surprising that Adam and Eve were said to have been pushed out of Paradise by the snake and the apple in the Bible. The apple is the symbol of excessive rationality—the exaggerated belief in knowledge—while the snake is the symbol of the ego.

3. The effects of wrong thoughts on health

Every illness starts from wrong thought patterns; in other words, they all have psychosomatic origins. If we live with a lot of negative thoughts for a long time and experience a lot of negative feelings due to them, our body will get ill sooner or later. The ego's favorite tool is to make negative emotions. This is because if we are scared for example, the ego can convince us easily that it's protecting us. This is the case with every negative feeling. Shame, guilt, apathy, fear, anger and especially pride have become embedded in our ego. The ego is strengthened by these negative emotions. The ego reaches these emotions with the thoughts that generate them. I used to think about future scenarios (needlessly) in which I had to protect myself or avoid other people's judgements and criticism. At the same time, I was continuously angry with things that happened differently from how I had imagined or planned. If something happened the way I'd planned, I was extremely proud of myself. My ego always put labels on other people and made comparisons. This pushed me into shame or guilt or made me proud. If I felt myself to be better at something than someone else, it boosted my pride. But if someone did something better than me, I felt shame or guilt. Do you think all of these negative feelings can disappear without a trace? Just look at most people over 40. Look at the condition of their bodies. Look at what they reflect in their faces. You may not feel it in your younger years, but it slowly uses up your health like a poison over the decades. In your twenties you ignore these; in your thirties you don't care about your mild symptoms. Then in your forties you start to regret why you've lived so wrongly. Health problems appear. They are different for everybody, depending on how many negative emotions they experienced from those mentioned.

4. The negative emotional effects of improper thoughts

As I wrote in the previous paragraph, improper thoughts geneate negative feelings. But also think about what chances you have for happiness alongside so many negative feelings. While I had my strong ego, I didn't understand why I wasn't happy, despite doing everything for it. I was hardworking, purposeful, and could handle the workload. I sacrificed myself to fight for my goals, but I was not really happy. How could I have been happy? I didn't realize how many negative feelings were inside me every day. I was used to it. What was more, my ego explained to me all the time that it wasn't my fault; it was the world's fault. My ego simply strengthened itself in me with it while actually destroying me. See how tricky it is?

5. Hubris

This is the trickiest part of ego. The ego makes us believe how cool, nice, smart and so on we are. Take an extremely appearance-conscious person for example who is the prisoner of their perfect body, or a rich, powerful entrepreneur. There are many other examples as well, of course. Hubris is what they have in common. The egos of both people make them believe that they are better than others. This feeling of being better gives them the feeling that they are special. Feeling special generates the feeling of happiness. The person with the perfect body shines in their vanity, enhanced by their appearance and by other people's compliments. Power and money can provide convenience and a sense of safety which give a sense of happiness to the entrepreneur mentioned above. Watching these as an outsider I see something else. I used to be an appearance perfectionist as well, and I am also an entrepreneur. The truth is that most of these people seem to be happy from the outside but there is often some emptiness inside. We ignore it and rush deeper into our perfectionism, lust for power or other addictive activity. Any excuse not to face the emptiness! Hubris gives just a shallow image of happiness which could be called a form of compensation from another perspective. The ego convinces us that with it we are on the right track. And ego convinces other people that they are the social examples and they are on the right track. This is the biggest trap and our whole social system is based on it. Just imagine what could happen if we cured the inner emptiness instead of the ego's hubris? What peace and harmony could fill our lives! This goal can be reached with an effort to get rid of the ego.

6. Distorted needs

A typical feature of the ego is that it always plans everything. If you plan even on your holiday what you will do and when, it means that unfortunately you have a strong ego. The other feature of the ego is that if something happens differently than you planned, it causes a big problem for you. Sometimes I am astonished at how big a fuss people can make at not getting their favorite coffee or ice cream. What are these problems while about one billion people can't get healthy drinking water, and three billion people are uncertain whether they will have food the following day? The ego overestimates itself and convinces you that your needs are really important. I'm not saying that they're not; I'm saying that the ego gives a distorted picture of your real needs. According to the ego, the world of emotions is not important, or the only important thing is to satisfy my own emotions. At the same time, the ego wants all the goods of the material world. It is only ME who counts. Why am I not worth a 600 m2 house and an SUV whose engine is as big as a truck's? It always has to be what I want—whatever it is. We don't realize how our ego distorts our needs. One of my acquaintances who became rich suddenly, said: 'I wish I could buy anything I want. For example, if I want to buy the twelve-hundredth gold watch I could buy it.' Our friendship ended with this sentence. I feel sorry for him but I can't do anything for him. His ego made him blind. And of course, here comes our ego's reaction: that our needs are more moderate, that is to say we're more normal. Be careful! Your ego always gives the next goal to you. If you reach your goal, you are happy about it but you start to think about the next one immediately. Your needs grow slowly from decade to decade. You will not realize that you have changed since they build within your personality slowly and gradually. A parasite also grows slowly in the host so as not to be rejected.

7. Lack of gut feelings

The ego is the constant planner—the pushy planner. The other aspect of this fact is that we become blind to our gut feelings and intuitions, however they are much wiser than our ego's advice. The ego convinces us that all intuition is bullshit. But it's just the opposite: the existence of the ego is bullshit. I'm not saying that the ego is unnecessary when I need to protect myself from a real mortal threat. Fortunately, this hardly ever happens in our life as there is so much peace and public safety around us. Fortunately, most people never experience a mortal threat in their lives.

8. The reduction of creativity

The fact that the ego plans systematically and thinks about the future all the time also kills creativity. The more callous one's ego is, the less creative that person is. This apparently destroys dramatically the chances of genuine efficiency.

9. The improper vision of safety

The ego constantly conjures up future hazards, theoretically, and keeps you in a permanent sense of danger so your body is in a survival mode all the time. It enhances stress and your body produces cortisol. Due to this, you never feel safe. How could you be healthy, happy and efficient that way? At the same time, the ego convinces you that you are efficient because this inner uncertainty keeps you alert all the time. This is just another tactic from parasites. If I feel safe, I can make decisions more carefully and make less trouble for myself.

Imagine if you lived without these, how much more peaceful, harmonious, efficient, healthy and happy your life would be! Be careful: your ego will make up millions of reasons to reject these words, but they are completely true. I have already destroyed a part of my ego, so I know it for sure. But when I had a strong ego, I would have ignored this book with some made-up reasons.

How do you destroy your ego? The answer is not easy. You have to evict a parasite that got into you when you were between 3 and 15 years old and slowly became ingrained in you. The first step in destroying your ego is to recognize its existence and to realize which areas of your life are strongly affected by it. You can find interesting writings about this topic in my blog. (The address of the website is at the beginning of the book.) In the blog section about ways of searching for happiness you can find several methods which can help you on this way.

APPENDIX 2

More about games

The word 'game' has been used many times in this book. I am sure most readers are aware of the meaning of this word in a psychological sense. It can also happen that others don't know what it means precisely. I will try not to bore you with simple definitions. I hope this appendix will be interesting to you even if you are aware of this issue.

Eric Berne, whose books I highly recommend to those who want to understand how the human spirit works, created the games theory. To those who want to learn about the basis to understanding how our spirit works and how to develop our self-awareness, I also highly recommend the book *Born to Win* (authors: Muriel James and Dorothy Jongeward), which uses simpler language.

When I first read *Games People Play* by Eric Berne in my early thirties, I found it really interesting, however I was convinced I didn't have games myself. It took about two years while the message of the book matured in me before I realized that I did also have one or two games. At that point I re-read the book and *Born to Win* as well. By the age of 36, I had a complete map about my games and I have to admit that I played a lot of different games then. At that time I came to know exactly when and what kind of game I was playing, and I also understood the spiritual roots of these games—their reasons from the past. These recognitions were crucial milestones in the development of my self-awareness. I have met a lot of people since then who know what games mean and can see other people's games clearly, but are not at all aware of their own.

Eric Berne uses the term 'transaction' for every form of communication between two or more people, as well as every action toward another person or people. Transactions are needed for the games, or we create games during our transactions. A game means that the goal visible on the surface during a transaction is different from the real one. This doesn't mean that we lie deliberately, but we want to reach something else than what we communicate during the transaction. There is a conversation for example between two people which has a target on the surface but below the surface there is another spiritual target that provides the motivation. This may seem complicated at first, but soon it will be clear.

To truly understand these games, we have to understand the real spiritual target first. For example, a typical game system for Hungarian people is complaining. I realized the big difference between Hungarian and American people when I lived in the USA. Hungarians complain all the time and walk in the streets with sad or neutral faces. Americans try to smile almost all the time and hardly ever complain. When an American starts complaining, it's because they need help. It's a straight, direct transaction without games: I complain because I want the other person to help me. So if the other person offers their help, the complaining person accepts it with joy and is grateful for the helping hand. In contrast, complaining is usually only a game in Hungary. We start to complain, but when the other person starts to give some advice with empathy or offers their help, we simply blow off them with something like 'oh, don't worry, it will work out'. This is only a game since the transaction on the surface is to ask for help but the real spiritual aim is to get confirmation of a highly negative worldview. We Hungarians like to believe that life is shit due to our history full of defeats over long generations, and we insist on this worldview. However, it is also true that some change in this attitude has begun in the past few decades.

Let's see another example. One of my games was that I always had to suffer much more for some success than others did, and everything was much harder for me than for others. It appeared in my communication in a way (unconsciously of course) that I managed my conversations to get confirmation regarding my worldview. When someone tried to point out that I could live simpler and easier, I just ignored it or tried to persuade them that my case was special. My case was not special of course; I just didn't want to depart from my worldview. The point of this game system is that we like to speak with people who give us this spiritual benefit. If someone doesn't want to take part in our games, we will avoid them next time and we decide that we really don't like those people.

So the real aim of these games is always to get self-confirmation. It doesn't matter how wrong, mistaken or self-destructive the worldview is; the aim is to collect the spiritual benefit of self-confirmation. Is it clear now why you can't just tell an alcoholic that what they're doing is wrong?

I used to know a lady who was absolutely honest in the sense that she could make love with someone only when true love and trust was given. In spite of this, her behavior with men was very coquettish and provocative. Her surface transactions made her seem to be a sexually easy woman. Those men who simply didn't care for her were evil in her opinion since they didn't pay attention to her. But she rejected those men in a rude way to whom her coquettish behavior appealed and captured their imagination. The aim of her transaction on the surface was 'come, woo me and you will get something wonderful', but the real aim of her transactions was to confirm her belief that 'every man wants the same thing, which is why you can't trust in them'.

All three examples show the same thing: we want to receive the confirmations to our thoughts spiritually. But if you observe these examples from the outside, you'll surely have some ideas of what kinds of childhood pain could cause such game systems and worldviews. So learning about your games will open up deeper ways to your self-awareness, which is the beginning of an exciting process. It can be stated anyway that the lower the level of consciousness one lives on, the more games they play in life, as they have to invest more energy in strengthening their worldview.

Most of my past relationships were ruined by my games, and it all happened without my being aware of it. Today, most of my relationships are free of games. I know clearly that they would never be perfectly free from games because then I would be a totally enlightened person. However, most of my relationships have become peaceful, harmonious and straightforward with the reduction in games. Due to this, my happiness has been enhanced. The very first step for everybody is to build into their self-awareness system the mapping of their whole game system. I wish You good luck in this journey.

Don't forget: your ego will start telling you that you don't have games, but don't let yourself be misled!

APPENDIX 3

More about the levels of consciousness

I wrote a brief summary of the levels of consciousness in order to make the basic notions understandable and to make it clear how and why people who are at the different levels have different perspectives about climate change. This appendix is for those who are interested in the levels in depth. There may be some overlapping with the text already written, but it serves to have this full explanation here in the appendix. Let's see the different levels in detail.

Shame and Guilt

Your level of consciousness depends on which level out of the 17 your spirit spends the most time at during an average day. Obviously, our states of mind are changing, because of which we're sometimes more enthusiastic and happier, and sometimes we're down in the dumps. Those people who live most of their lives at the level of Shame and Guilt have the lowest level of consciousness. According to the scale fixed by David R. Hawkins' measurements, the spirits of people who live at the level of Shame have their energetic level at about 20, while people living at the level of Guilt have an energetic level of about 30. Just to understand, the level of Death is 0 and the highest level is Enlightenment, whose energetic level is 1000. The Life-destructive levels are below 200, while Life-supportive levels are above 200. It is a scientifically proved logarithmic scale about which you can learn more in David R. Hawkins' book. I don't want to bore anyone with scientific details, so let's talk about the more interesting practical part.

To recall and re-experience these feelings in my life brought me closer to understanding people. How many people there must be in the world who have suppressed similar tortures within themselves, which they carry deep in their spirits and make them play their craziest games unconsciously! Because of childhood traumas, loss of parents, a painful divorce, lack of attention because of another sibling, parents destroying the children's souls, deprivation, exploitation or other reasons, these deep spiritual wounds burden billions of people. It gave me the recognition that we can't judge other people since we can never know why the other person behaves the way they do. What's more exciting is that if I had been in that person's life, I would have done the same. Jesus is said to have referred to it when he said to forgive the people who act against us. They do it unconsciously to suppress their hurt spirit. If I hit back, I will just deepen their wound. If I forgive them, I help them in healing their spirit. We can't apparently fulfill these words below a certain level of consciousness. Since we have so many spiritual burdens on our shoulders, we do not have the capacity to deal with others and we react to the smallest attack with great resistance and suddenness. Regular forgiveness starts above the Life-supportive level of 200.

Apathy and Grief

The energetic level of Apathy is 50 and of Grief is 75. These are much higher than the levels of Shame and Guilt. It can be surprising, but being sad or apathetic is at a lot less negative level than the levels of Shame and Guilt. I experienced these levels for a long time in my life and I do not wish it on anybody. But those who are at that level also have a chance to leave it. However, if we saw the world from the perspective of Shame and Guilt, it is a big step forward. It is Apathy and then Grief which signify the escape from the lower spiritual levels. This is why it's wrong to judge an apathetic person. That person may have just stepped up to a higher level, but as you judge them through your filter (which is of course not accurate), you simply can't imagine that their low level can be a development. It can happen that this person moved from the level of Guilt to the level of Apathy. This would be a huge development, and by judging them we risk pushing them back to the lower levels. These people must be left in peace to let their spirits rest a bit at this level, and some inspiring power will come which can move them forward. This is the time when you have to help them, but not with judgment! It can also happen that they've fallen back from a higher level to this level. We have to give time to ourselves at the level of Grief. We have to experience the pain of our grief, and that will help us gain energy to move back to a higher level again.

It can also happen that the pain of the grief tortures us so much that it pushes us back to the level of Apathy where our spirit can rest a bit before it starts to develop to higher levels. Time must be allowed for both levels. Those who run away and hide their grief will have troubles later in their lives. On the other hand, if the levels of Apathy and Grief stay permanent, then it's time to ask for external help! This level of consciousness is so deep that it is difficult to move on from there without help.

People who live at the level of Apathy are not interested in anything, and nothing matters to them. They feel this world can give them nothing valuable. There is total hopelessness, but at least shame and guilt do not torture the spirit. The sense of death often appears here as well as the need for external help, but we do not have the power or mood to find a helper. This is why it's good to have a good family and good friends who can offer a helping hand. As our future vision is full of desperation and hopelessness, we see the future very pessimistically and the world dark when in this state. Climate change means destruction for these people. We give up the positive things approaching us or we simply do not care about them. Because of the negative feeling that fills our spirit, we condemn our environment and can refuse help.

The level of Grief is a progress compared to apathy. Although it is still a very low level, some spiritual activity appears, as we have active emotions here even if they are painful. We do not feel pain at the level of Apathy any more, or if so, we don't care about it. At the level of Grief a desire appears for the pain to go away. This will help us to move to a higher level. Of course, this moving on will be healthy and permanent if we let the grief, coming from mourning or another reason, pass through us and we **experience the spiritual pain and do not run away from it.** This heals the spiritual problem, not just hides it. Our future vision at the level of grief is still really dark. We are discouraged spiritually and are likely to be stuck in the depths of self-pity. We consider our life to be tragic and react to help disdainfully, thinking 'they don't know what I'm going through'. This is the state when we nurse our spiritual wounds, and it will help in healing them if we do it long enough. In this state we don't have the capacity to do anything for climate change as we aren't able to do anything for our own life. But these people can't be judged for it; if you were them, you wouldn't have the capacity either. This is why we have to help these people move to a higher level of happiness.

Fear

All of us have been scared at some point, which is a natural reaction of our spirit. It is healthy and right in itself. Just because we are scared sometimes it doesn't mean that we are at that level of consciousness, as we have to take the average of all our feelings. If we want to observe our self-awareness from the perspective of where we can be on that scale, we have to approach it this way: our spirit is at the level of fear if most of our thoughts are controlled by fear. I experienced this level permanently as well, and I do not wish it on anyone either. The energetic value of this level is 100, so it can be seen that it is above the level of Grief.

People living at the level of Fear are not scared of one certain thing; they are likely to be scared in general. For them, the whole world is really scary and they hardly ever feel safe. Of course, every such person has certain things they are scared of. Due to this, a person living at the level of Fear is anxious. In company and every other place where they don't feel safe, their behavior is reserved. Sometimes they're just the opposite and whenever they can they enthusiastically talk about the reasons for their fears. These people's future visions are strongly pessimistic regarding both their own lives and Humankind's future. They are the people who wear face masks even in deserted streets due to the pandemic situation, and they are the ones who see the instant collapse of the world as a result of ominous climate news. It often happens that the fear of sexually transmitted diseases and pregnancy control their spirit so much that they keep away from sexuality, or if they do make love they cannot feel free during the act. These People feel threatened in almost every situation. Their ego always screens the future for where they will be attacked from. As a result, their spirit is in constant standby and they hardly ever dare to be free. They step out of their calmness quickly, lose their temper and do things they later regret. They tend to use mind-altering substances (alcohol, drugs) since they can experience freedom only in that way. Religious people at this level imagine God as a punisher who deals with them and others harshly because of their sins. This further boosts their fear. If these people can have control over others, they become punishers as well. These people have a lot of Life-destructive energy unconsciously because of these features, but less than people living at lower levels.

To reach this level from the levels of Shame or Guilt, Grief or Apathy represent great progress. The individual's spiritual activity is much higher at this level. We should never look down on people living in fear as it can happen that they had just moved on to that level, which was a definitely big progress in their life. We should support and encourage them to leave this level for a happier and higher one by gaining power from our help.

Of course, we can fall back from higher levels to this one and it's really hard to get back. I recommend that these people ask for external help, although most people feel ashamed to talk about their fears. Indeed, people living at the level of Fear are often scared of talking about their fears.

Unfortunately, making people scared has historically been a tool of political powers and churches, and it is still so. That is to say, people pushed to live in fear can be easily controlled and managed. There are still political or other powers and religious sects where this tool is used extensively. These systems cannot survive for a long time because people instinctively want to live happily. And this level is very far from happiness and respect for Life.

Desire

The energetic level of Desire is 125. As I have already mentioned, the levels below 200 are all Life-destructive, so this level is still pretty negative and degrading to both the person and their environment. It is true however, that compared to Fear (or other lower) levels, people reach a higher level here if their spirit stays at this level permanently. If we observe the levels from Shame to Fear and see the level of Desire from that perspective, we can find some more effective spiritual activity here. As this level is not so low, hope appears as a strong feeling and our spirit starts to long for something good. This level can act as a springboard to the higher levels. Desire gives us strength to start for our targets. At the same time, the level of Desire is really negative when seen from the higher levels. This level is the world of addictions. I can say that those who live at this level permanently are suffering from addictions. The basic feeling of addictions is longing and we are so used to the hormonal tingling generated by desire that what is important is not to reach the target but to maintain the desire. This ill state of mind is the reason why addicted people do not want to admit that their addiction creates problems and that they should give up their addictions. At this level we insist on longing so much that we think our life would be empty, plain and boring without it. Those who are addicted to porn movies love longing for the sexual targets seen in the movies and when they can finally experience it live, they don't find it as good as they imagined. While experiencing it in reality, they lose the taste of longing and in that way their orgasm can fail as well. The porn industry is a good example for this, but we can say in general that one of the main driving forces of the current economic system and how society works is to make people crave. Deals, exciting products and services are offered through advertisements and other media for which people can yearn.

This social system causes an incredible number of people in the world to live at that level. It's really easy to find targets we can long for and as we have a strong tendency to do at on this level, society can achieve its objective and provoke people to consume. Addiction to porn and sex is characteristic mainly for men due to hormonal and other reasons. I know a lot of people who judge men for these problems and call it disgusting while not recognizing their own addiction to shopping or other things. A typical common feature of the many forms of addiction forms is that our addiction seems normal to us because our ego convinces us that it is right. Other people's addictions are reprehensible and disgusting. However, addiction is addiction, with the same spiritual root and the same negative level. These people must not be judged but they must be supported to reach a higher level of consciousness. By judging them we just keep them at this level, as due to the new disappointment they will escape into their addictions.

Those who stay at this level and give up one of their addictions (because of its harmful effects) will choose another one instinctively. I know people who gave up smoking from one day to the other but have become slaves to other passions.

The mindset of people who live at that level is about disappointments. It sounds pretty logical because if there were no disappointments in their life, they would reach their targets sooner or later and there would be no reason for their longing. I also stayed at this level for a long time and experienced a lot of disappointments as well. I always started for my goals with renewed energy, then they collapsed and I had to start everything again from the beginning. If I reached my objectives in a certain area of my life, other areas collapsed. This way I could always long for other unfulfilled targets.

The disappointed mindset is often followed by a mindset that rejects God. These People think if there were a God, the world would be different. They do not admit, because they do not know, that they are the causes of the lot of their own disappointments. They think that Life is against them and they are ill-fated. Apparently, the solution could be to realize that we are the reasons for all of our troubles.

These people start out for their unreachable targets with enormous force, and the failure and disappointment will be there at the end. Their conscious self longs for the target in vain, and their unconscious self searches for the failure because they need disappointment spiritually. That's why in this energetic system what the spirit delivers towards its environment and itself is so Life-destructive. The other main reason is the harmful impact that addictions have on their environment and on the person.

Anger

The energetic value of this level is 150, which means that we are still talking about the Life-destructive levels below 200. From the perspective of the other levels, here we can find real activity. The individual is negative both toward themselves and their environment, however, the individual already expresses strong self-defense and delivers strong energies. One of the main features of people living at this level is that they always blame others and they express it with harsh criticism. They judge everybody who sees the world differently from them and who reacts differently from what they expect. Empathy is really low at this level. We have all been angry at some point, so we clearly know this feeling well. Those who live at this level feel like this during most of their life. A good example for this level is Grouchy from the Smurfs, although he's presented in a lovable and more solid way. It works in a much more destructive way in real life. Hate is the main feeling that steeps the spirit. People living at this level always search for a target person or entity they can hate. These people easily get into fights or other aggressive situations. They are the troublemakers at clubs. These are the people who always litigate with someone and they usually lose the case so they can hate the judge as well. The circle of hate never stops. Those climate activists who try to bully the world into changing with violence, fights and rude words are also at this level. Obviously, they are also right in the sense that we have to sway the world out of its current wrong direction.

The main focus of these people is the feeling of hate even if they communicate just the opposite. The individual always searches for confrontation unconsciously. They are convinced that the world is full of evil things and wickedness, and the most part of it is against them. In case if they are believers, they are convinced that God is a vindictive entity who avenges every bad action.

If someone lives at a lower level, Anger could be the next step in their development. This is the first level where we can strongly stand up for ourselves. People living at lower levels instead struggle within themselves: they can hardly ever defend their interests. They usually obey, withdraw, adapt, or capitulate, but at this level confrontation appears and as a result the individual stands up for their interests.

On the other hand, if we fall here from a higher level, there are real dangers that we could get stuck here. We can sink here for a short time to freak out and then start for a higher level of consciousness with renewed energy. If we see it from that perspective, it's necessary to freak out to be able to move on to a higher level. Suppression cannot be a solution at all; it keeps us at this level permanently or pushes us even lower. If you are stuck at this level, it's suggested that you ask for help. The real reason for the opposition with the world is generally not an external reason but something inside us. Most of the time we project to the world that we are dissatisfied and angry with ourselves. It's much easier to blame others and be angry with others than to face and reorganize ourselves. If you have lived at this level and was able to leave it, then you know what I mean. If you recognize yourself and think you live on this level, it can happen that you're reading these lines to move on to higher, happier levels. Anger mostly destroys us and burns us inside so it can be the source of serious mental and physical diseases. This is why it's important to develop and move on to higher levels.

Anger is one of the basic emotions which we are always able to feel, as it is one of our innate emotions. Anger, shame, hate, disgust, fright, disappointment, joy, love and curiosity are the basic emotions according to Paul Ekman in his book *Emotions Revealed*, which we don't have to learn from others in relationships. A human is an open material and energy system, so it needs the continuous supply and flow from both. Energy is superior to material, even if we think the opposite due to our current materialistic view. If there is no energy, there is no avatar, no material reaction, no activation energy. Anger is a final activation energy we can use unconsciously or deliberately anytime to have strength, to have the power to confront, to stand up for ourselves, or just to have energy at all. This is the reason why so many of us are stuck at the level of Anger.

People stuck at this level do not dare to move on from this level because they are afraid they would lose the excitement of confrontation and their life would become empty and boring. There is no need to be concerned about this! Life is happier at higher levels, not more boring. They are also scared of the fact that if they are not angry with others, they will have to face themselves and they wouldn't dare to do it. They are scared of what they would find in themselves, so to hate others is nothing else but to escape from themselves. However, what you find inside is the key to your happiness! You have to be brave to look inside.

Pride

The energetic value of this level is 175; this is the last Life-destructive level.

However, the ego is deeply present in the spirit's structure on every Life-destructive level, and this is the level where the ego gets the strongest space. The emotional attitude of people living at this level tries to separate them from others. They pretend to seem to be more special and better than the judged target group or person. The main direction of emotion is minimization. We minimize others to seem better. Conceit and being puffed-up are parts of this emotional attitude. People living at this level can disguise their infinite selfishness pretty well if they are intelligent enough. Their selfish interest doesn't seem selfish, so in spite of the Life-destructive level, it is often the case that we see a friendly and observant person on the surface. Seeing behind the selfish outer shelf is possible only with knowing the person deeper.

A lot of relationships fail because later it turns out that the partner who seems to be pleasant at the beginning is unable to be selfless, which is the basis for confident relationships.

Most people live at this since the world inspires us to be selfish and deal only with ourselves. Since it is still a Life-destructive level, unprecedented environment destruction and ominous climate change forecasts are revealed. Due to the 'blessed' activities of internet and media, most teenagers and young people live at this level as well. As a result, real communities fall apart in the younger generations, the existence of real community is about to disappear and the average of divorces is above 50%. Everybody deals with themselves and expects everyone to deal with them. It is quite hopeless to make real and deep relationships. It is of course always the other person because of whom the relationship fails. The individual does not realize the fact on this level that the more selfish they are, the further they get from having real and quality relationships, not to mention that they destroy natural resources more as well. A selfish person wolfs down and experiences everything they want. They don't have enough self-control and they don't need it. The aim on this level is to experience what they can, reach the desired targets, possess material goods, and have everything that one can have.

A person who lives at the level of Pride can do everything only because of self-interest. Their only motivation is themselves and to reach their own goals. They are not able to love selflessly. They love other people to receive compensation. Furthermore, these people are convinced that there is no selflessness, and that selfless people behave that way just because they know they will benefit from being selfless. These people can be identified with the saying 'every man kindles the fire below his own pot'. When I lived at that level, I was also convinced about it and didn't understand those who talked about real selflessness. I thought I was right, and they only idealized the notion of selflessness. The truth is that if someone living on the level of Pride has never been on a higher level, they can't believe there is real selflessness as they've never experienced it, therefore they can't be judged because of their narrowed perspective. It is as if someone has watched the sky from the bottom of a steep ravine during their whole life and we want to explain to them what a sunrise is like.

A lot of people who are successful in their job, financial or political situation live at this level. This success can be misleading and it's difficult to imagine that they deliver Life-destructive energy to both their environment and themselves, but it's the truth. The root of Life-destructive energy is that people are not ready to prioritize social, ecological or other public interests, and as a result of their selfishness they forget about the real aim of their life. They can identify with the real aims only if they are in accordance with their own, so it is pretty clear that they would identify with those views and ideas which can support their own interests. They always believe in things which enhance their self-interest and self-image and want to convince the people in their environment about it. We behave the same way with everything that builds in the ego. Our consciousness narrows down dramatically without recognizing it. Furthermore, in most cases we clearly know and believe that we are smarter, nicer and better than others. These people are absolutely aware of the fact that they are worth their targets. They do everything to reach their targets, they are also ready to break the rules. They are really demanding, which means they want their due in every area of life. They hide it perfectly until they are on the way to reach their targets. During this time they behave politely and they are calm on the surface. But when there is some uncertainty about reaching their target, they become really demanding and show their real colors.

These people are mostly atheists or indifferent to the idea of God. The image of God delivers selflessness and all the world's monotheistic religions teach this. It is not in accordance with the selfish personality and its aims though. Most of these people think that the basic religious concepts are not for them. They may follow some moral rules, but if they break others, they explain to themselves why it was an exception and why it was right. The ego, persuading itself, is beyond every rule and helps the individual not to have any remorse about it. If the individual had some remorse, the ego would explain the problem and suppress it.

I spent long years at that level of consciousness. I was convinced about my truth and I did not recognize in what a Life-destructive way

I lived. My narrow-mindedness was really distorted, however I thought just the opposite about myself. I thought I was on the right track, but now I know that selfishness is really far from the right way if we are searching for real happiness. Professional, financial or other successes make us blind at that level. Our ego makes us believe what we are doing is right since we can think more of ourselves that way. The solution for leaving this level is to look behind our ego and see through its mirage. Only a few people are able to do that, so most of them are stuck at this level. However, the really successful and happy people have already moved on and have reached higher levels.

If we look at this level from a higher level of consciousness, we will see a narrow-minded and selfish person. If we fall back to it from a higher level, the spirit can rest here for a short time and deal with itself only temporarily. As long we are fighting with the dragons in our spirit, we do not have time for selflessness. The fight with our inner spiritual problems uses so much energy that this is the reason why our vision narrows down and why we become selfish.

However, if we move higher we can reach only Life-supportive levels, so this is the last level above which real value creation, real success and real happiness are waiting for us. This motivation is enough to step out from the world of self-deception to a more valuable level—one which provides real Life.

But if we observe the level of Pride from the level of Anger or from lower levels, it represents serious progress and development. The individual here uses a much more sophisticated system to express itself. The driving force of the ego is not the fear of others or confrontation with others any more, but 'only' the separation from others. It is not so aggressive, and it is less environment- and self-destructive. It's no surprise that this level is the highest among the Life-destructive ones.

On the other hand, to step out of this level and to move on is the most difficult, since our ego convinces us that everything is all right because we are different from the other people to whom we compare ourselves. We feel better, smarter and more beautiful. Our unconscious self compares us to whom we can have advantage over. If we get to know people who are obviously better than us, we try to stay close to them and pretend we are better by being their acquaintances. The other strategy is to make up reasons why we don't want to be so good. As this level is a self-deceiving mirage, it misleads people and they don't want to move on to higher levels. This is a group for unconscious losers, while the winners experience themselves consciously.

It is important to know that people living at the level of Pride are always lonely inside. Selfishness separates the individual deep in the spirit since the individual doesn't believe in the power of unity. A person living on this level is scared deep inside. The root of the fear is the fear of unity with others and dependence on others. We do not want to depend on others because we are scared that we will be hurt. We don't have the courage to love people more than they love us because we are scared they will take advantage of it. Our ego always observes who returns what we give them and to what extent, and upon the smallest sign that they are not behaving as expected, we withdraw into our separated, selfish world. From that perspective it can be clearly seen that selfishness is a self-defense armor which we grow because of childhood grudges or other hurts. As with many other Life-destructive levels, it also feeds from the deeply suppressed root of fear.

Courage

The level of courage is where the spirit rises to a Life-supportive level (its energetic value is 200). When people step from the level of Pride to the level of Courage, a quality change appears in their life. It appears gradually of course, as it takes a long time to solve the problems of the past.

Courage should not be confused with audacity. Audacity is a symptom of Life-destructive levels. A bold person can seem brave from several perspectives (they are cornered so they fight back, or they don't take consequences into consideration etc.). People living at the level of Courage are not recognized for how they defy their enemies. On the contrary, these people dare to accept their weaknesses, mistakes and faults.

A student told me about a teacher recently who admitted that he didn't understand the solving method that the three smartest students figured out on his lesson. He taught an easier method. Surprisingly, students still look up to this teacher; they still like him. This is not by coincidence, since it can be seen from his gesture that this teacher lives at least at the level of Courage. He admitted in front of his own students that he is less smart than some of them. At the same time, he has authority and is popular because that's how Life-supportive levels impact people in their environment. It's the first level where the individual faces themselves honestly (compared to the previously introduced levels). This is the most important change compared to the previous levels: people at this level are not afraid of showing real mirrors to themselves. They start to see their mistakes, weaknesses and frailties realistically. Furthermore, they are able to open up fears and pains stored and suppressed in their subconscious. This is why Courage is the first Life-supportive level, as we start to have a realistic self-awareness and worldview here. I must note here that every person living on Life-destructive levels thinks that they have good self-awareness. We start to doubt this only at higher levels. Here we are on a path where we do not escape from ourselves and we accept ourselves as we are. Here we start to take on ourselves in front of the world with all of our good and bad habits, and we start to free ourselves from the dragons in our spirit, which caused so many bad things while facing our pains. A huge change at this level is that the individual finally takes full responsibility for their deeds and doesn't blame others.

A basic feature of successful people as opposed to losers is that they take responsibility for their deeds and if it is negative, they learn from it, look ahead, and do not hide from the burden of responsibility by blaming others.

This is why the unlucky person can be lucky, the unsuccessful successful, the selfish opens to selflessness, and we start to give love in a way that we don't expect any in return. All of these things are not routine at this level of course, but at least they appear in our life and have a positive impact on us. As a result, the spirit can defeat its dragons and is able to move in a more free and happy direction.

Now we finally sometimes have the capacity to do things which develop our spirit, and we will more frequently have the need to look after others selflessly. Now we start to do things for our real life tasks, stepping out of the struggling world of necessities and daily fights. Obviously, I'm not talking about being able to buy a better car. However, success can be seen in our more stable financial situation. My financial situation has always been really fluctuating.

If somebody has just stepped up to this level in their spiritual development from a lower level, they are only at the door of happiness. In order to step in and stay there permanently, we have to work hard on the hardships which come from facing ourselves. It's really hard work to free ourselves from our inner fears and pains, and it generally takes a lot of time. People who want to develop therefore stay here for a long time, but people who skip it can experience an accelerating spiritual development in the direction of even higher Life-supportive levels and a higher level of happiness.

Those people who fall back from a higher level are here to gather courage to overcome another spiritual problem coming from the subconscious to be able to move on to higher levels than before the fall.

People's main emotion who live at the level of Courage is confirmation. They encourage and confirm themselves and others. Their main objective is to be able to achieve their goals. They place their development at the service of reaching these skills. If they succeed, they will move on to higher levels. These people take an active part in fulfilling their wishes without doing harm to others, presenting them negatively, exploiting or cheating others but instead developing themselves. This is one way among others in which they fundamentally differ from people at the level of Pride or lower levels. If these people are believers, they imagine God as a forgiving entity whose love is for everybody, including those who make mistakes or are feeble sometimes. The frightening tactic of religions or the demagogic influencing of sects have no impact on these people.

Neutrality

If we step up in our spiritual development, we reach the level of Neutrality. The energetic value of this level is 250. The individual has already stepped into the gate of happiness on this level and things are already stable in their life. The main motif of their life is to experience the positive things. People living at this level can reach their successes without any special effort and more importantly without damaging anyone. Excessive self-assertion does not exist here. People at the level of Pride are also often successful, but to exploit, minimize or damage others is somehow always part of their success. This behavior does not even appear at the level of Neutrality. Watching these people from lower levels, we usually admire them and wonder: how can they do it? A long time ago I didn't understand either how they can proceed step by step without any fight or struggle. At lower levels, Life is about the everyday struggles, but at this level the experiencing of Life in a positive and peaceful way emerges. To fight is important at this level, but stressing about problems is not really characteristic.

The level of Neutrality is the first level where we are relatively satisfied with our life and ourselves without lying to ourselves. The same happens at the level of Pride but with lying. This is a really big thing since we have not talked about any levels where it's true. When I arrived at this level, my vision about the world and the future and my approach to people changed suddenly; I started to feel some really deep and calming peace and harmony inside. This kind of sense of spiritual safety had been totally unknown to me before. My future vision also changed from pessimistic to optimistic. I wish everybody could live at this level! If it were so, there would surely be world peace and environment pollution would fall to a very low level.

If a person living on this level is a believer, they imagine that God gave them their unique abilities to be able to benefit others. I realized at this level that I was meant to be a teacher and that this was the special ability I got from Life. Since I knew it for sure, I have been giving my lessons at the university with more compassion and conscientiousness. It also inspired me to write this book.

At this level, the main development process of our spirit is release. Here we get rid of the dragons in our spirit and the addictions which pulled us down before. The release, however, is not complex here; we can still develop further but we can quit a lot of the habits, games and addictions forever which pulled us down to the lower levels. It causes such a great feeling of freedom in us that releases a lot of energy immediately. Up to this level we had always struggled with our inner problems which had taken a lot of energy consciously or unconsciously, and we had also struggled with the consequences of the bad decisions we made instinctively. Here our freed spirit becomes easier and starts in the directions of our main tasks. Energy is finally used for Life-supportive and selfless actions. The writing of this book is also Life-supportive and selfless, but it is still a very good feeling since the lot of positive energy wants to break out from me and wants to do something to make the world better.

There is another breakthrough at this level, namely: full trust becomes an integral part of Life. A leader at the level of Pride doesn't trust in any of their employees and wants to control everything. A leader at the level of Neutrality trusts in the success of interdependence and reaches bigger successes by it. Trusting in my teammate's decisions is the basis for managing high-level teamwork. This success cannot be measured in profits but by how useful the team members are socially or in their spiritual and professional development. People living at this level are not jealous in their relationship and they are able to fully trust in their partner's decisions. It's important to highlight that letting my partner do what they want and trusting in them are far from being the same thing. The individual starts to trust in people's goodness and in a positive future. It's also a new thing since neither a strong trust in the future nor trust in people's goodness were characteristic of the previous levels.

These people as seen from lower levels can be said to be naive or idealist since on lower levels it seems impossible that optimism can be objective and well-established, but it can!

It's a wonderful level of consciousness. Most of the people living today can never reach it, unfortunately. Why? The answer is simple:

because most people spend their life dealing with formalities instead of their spiritual development. It's not bad chasing after money, a career or a beautiful appearance but these do not lead to happiness at all. But this level really is the level of happy people. As I joined this level a few years ago, I clearly know how good it is and I would really like you to experience it if you haven't so far. With it, you can do the most good for the world's future beyond your own happiness!

Willingness

I am really happy that you have reached so far in reading this appendix about the levels of consciousness. The level I am writing about now and all the following ones are the real platforms of happiness. There are differences among these levels only in the experience and stability of happiness.

The energetic value of the level of Willingness is 310, so we are high above the level of 200 which is the limit between Life-destructive and Life-supportive levels. As of this writing, it has been less than 3 years since I reached this level and my life changed into a higher standard of living. People living at the level of Willingness differ mainly from the other levels in that they are basically optimistic. These people can hardly be swayed from their optimistic approach. It becomes instinctive to see the glass half full and if something bad happens, we draw the positive conclusions. People have a so-called 'good time' inside, which means that a basic optimism characterizes their life regardless the external happenings. Of course, this doesn't mean that People living at this level can't feel low or be nervous, but it happens rarely and only for a short time.

There is a strong intention in the individual to develop further in the spiritual world. At this level we want to abandon every inner process that causes negative energy deliberately. It becomes obvious that happiness comes from inside and it is almost totally independent of external happenings. Now we see what those spiritual leaders mean who talk about it all the time. I have swallowed recognition from other people my whole Life to fill the lack of self-confidence and self-acceptance suppressed in my spirit. Unfortunately, this was not successful. However, the many recognitions embedded in the ego may provide some kind of complacency. On the other hand, this gives no real happiness as it is based on cheating ourselves. We don't accept ourselves deep inside; we just believe we are valuable due to other people's feedback. When my spirit changed inside, my self-acceptance developed and I started to love myself. My worldview became positive immediately without the need for external confirmation. It is an actual and real solution on the way to happiness.

At this level our worldview is hopeful and optimistic. The individual's worldview about the future is characterized by problem solving and trust in Life (or in case of believers, God). If the individual is a believer, they imagine God as a motivating entity, who motivates them to use their talent to change the world positively.

If someone comes to this level from a higher one, the intention arisen here will give the power to move on to higher levels again. During my spiritual work at this level, I worked on releasing such deep spiritual repressions and problems about which I hadn't had the faintest idea. Spiritual development works like this: when we reveal a layer and sort out the upcoming spiritual problems, other layers appear which were hidden before. This is true of every level of spiritual development of course. I am astonished by the fact that after 18 years of conscious spiritual development, there are still problems to solve. This represents well the infinite diversity of the human spirit, from which the more you experience, the more gates will open up to you.

If someone arrives from lower levels, they experience a new dimension of self-acceptance which brings an inner balance to their life. The individual understands and feels at this level why internal peace is the basis for happiness and not the chasing of positive experiences. It is balance that matters here, and not the constant search for adventures. This is the particular balance that is the key to a solution for climate change and steeps this book.

When a spiritual leader talks or writes that it's no good to be crazy in love with somebody or escape from everything consciously and bolt down good experiences, most people just don't understand it. Furthermore, their reaction can be hostile because everybody wants to be crazy in love with somebody, and everybody wants to have good experiences. At this level, people can understand this message. The spirit can find balance here, which results in a wonderful internal tranquility. If our spirit sways either in a positive or in negative direction, it wants to get back here immediately. It's like the difference between wavy and glassy water. The lower the level of consciousness someone lives at, the more wavy and heavy water their spiritual processes can be compared to. Emotional extremities are big at lower levels; positive waves are followed by deep negative ones. At this level, the spirit is like glassy water which starts to wave easily due to external impacts. But the surface becomes glassy quickly again because the spirit searches for this state instinctively. This is boring and monotonous to somebody living on lower levels. They may see the harmony and peace in this state, but there are no happenings. They need to love passionately and fall deeply to struggle afterwards. They need sparkle in their life. This is natural since these sufferings will move them on in their spiritual development. Their spirit is like glassy water only for short periods when their spirit is too tired to sway.

One of the biggest mistakes of western society is that we try to separate ourselves from negative things and we want to secure all the good things Life can give. Meanwhile, we forget about our spiritual development, which often results in tough, long-term torture. Why? Because after a permanent positive wave an even more permanent negative wave comes. The spirit is also an energy system, so the basic rules of physics work here. But I would like to write about this in detail in another book.

Acceptance

The level of Acceptance is a really wonderful and happy world. In the last two years I have stayed here during my spiritual development. Starting from the level of Shame and Guilt, I have reached this level in 18 years with more serious fallbacks. This and my subsequent books are written to help you to make this development shorter, meaning that you can go along a beaten path without making detours in your development.

The level of Acceptance is the most efficient at the energetic value of 350. It is deep inside the Life-supportive range. As the increasing of the levels is not linear (but logarithmic), it is such a high level that it can neutralize the negative energy of 10,000 Life-destructive people. That's why when you listen to a lecture where the lecturer is such a person, the whole lecture holds you, you don't realize the time passing by, and your stress and tension disappear. The interaction of levels of consciousness work like this.

This is the first level where the individual accepts themselves as they are. And it is not because of self-suggestion or resignation. By self-suggestion I mean when we persuade and explain to ourselves that we are good the way we are. By resignation I mean when the individual feels they cannot change and becomes uninterested in themselves. Self-acceptance at this level is an inner, honest feeling. It is not about being conceited since we see our positive and negative features realistically. We love our positive values and accept our negative ones. I was running away from myself almost my whole life. How I got here is a wonder to me. I have never thought it was possible. Accepting ourselves brings inner peace and harmony. It's like when the spirit arrives home finally. Here we start to find ourselves and live according to our life's tasks. Accepting our negative features does not mean that we don't want to change them and become better, of course. We still want to develop ourselves at this level, but without beating ourselves up because of the bad things in us. Internal frustrations disappear here.

Accepting ourselves results in accepting our environment. If our level of consciousness changes inside, our perspective changes as well. We become forgiving of people in our environment and accept the world as it is. Of course, this doesn't mean that we don't want to make our environment better either. It means that we can use the energy released from the lack of redundant opposition to change the world positively. I am frequently asked: 'How do you have the energy for this book besides so many other things?' The answer is simple: all of my energy that I used to use for the fight against myself and the world has been released. I used to suffer from a lack of fortitude all the time. Nowadays I feel it only once to twice a year and only for a few hours. Our inner fights burn an incredible amount of energy when we're at lower levels. But at this level, harmony is our worldview that steeps our way of life, so there are no inner fights.

The other main emotion at this level besides acceptance is forgiving. We forgive ourselves all the sins and bad deeds of our life, and we do everything to make them better in the future. We realize at this level how many bad things we've done so far. At the level of Pride (or other Life-destructive levels) we don't realize how many bad things we do. We destroy our environment unconsciously while we're convinced about our personal truth. At those levels, our ego hides the consequences of our bad deeds from us and explains to us why what we did was right. At the level of Acceptance we truly see and feel how many bad things we've done and do. Here we take all the consequences of our deeds and we do everything we can to make them better.

Forgiving works with other people as well. Either we are not really angry with people doing bad things to us, or we are angry only for a short time because we know well what spiritual dragons they're fighting with and that they have only hurt us unconsciously. We can help them if they let us, but usually they're at the level of opposition so they refuse the help, and in most cases they aren't even able to understand it. They make up some ulterior motif because they can't believe that an 'enemy' wants to help them selflessly. It is beyond comprehension for them. Anyway, Jesus' words make sense on this level: 'as we forgive those who trespass against us'.

Our spirit starts to become transcendent on this level. The narrow-mindedness and borders of rationality are sharply outlined here. I highly respect rationality as an engineer, researcher and university lecturer, but now I know that the world of rationality is really tight. Something can exist even though we can't describe it in a mathematical-physical context. Furthermore, if we reverse this idea, we hardly know any existing thing so well that we are able to fully describe it in a mathematical-physical context. Just think of love or its counterpart, hate. Everyone knows they exist but they are not touchable rationally.

If someone believes in God at this level, they see God as a merciful entity who forgives their sins and supports us to make them better. Supporting Life means here that we take an active part in making all the things better we that have done wrong in our life. As our internal tensions are replaced by harmony, we have the capacity and fortitude to take an active part in it. If 20% of people reached this level, there would be world peace and the environment pollution would reduce to such a low level that the destruction of the environment would stop and the environment would start to regenerate globally. As I have already experienced this way, I know it's possible! You can step onto this path too!

Reason

I have a close friend who has been living at this level for a long time. I have always admired him for it. This level can be characterized by the energetic value of 400 and represents a very high Life-supportive level. Einstein and Newton lived at this level as well. If this level is accompanied by high cognitive abilities, it brings outstanding performance. My friend also does his job on a very high level even if not on the same level as the mentioned scientists. To do our profession on a very high level is necessarily part of this level.

The individual reaches the world of objectivity on this level. A lot of people think they are on this level, but only about one in a million people can reach it. The reason is that people are convinced about their objectivity but 99.99% of people are far from being objective. Their feelings and narrow mindset distort the individual's perception at lower levels. Most of the time this happens unconsciously. People are more rationalizing than rational at this level. It means that the decision has been already made, based on emotions or experiences, and the individual tries to support it by their rational thoughts afterwards, while on the surface they think they made their decisions rationally. The professional advertisements affecting our emotions make use of this, and this is the reason why these advertisements can work better in case of people living at low levels. I had always been proud of myself that advertisements didn't have an impact on me because I distanced myself from them as much as I could. This was totally wrong 'objectivity' at that time. The reason I distanced myself from the advertisements was that they actually had an enormous impact on me and it disturbed me. This is a good example of how we rationalize things which we conceitedly think were totally objective afterwards, and we are fully convinced about our truth. The level of Courage is where we dare to see behind this.

At the level of Reason the individual is able to appreciate and analyze the information they sense, fully objectively, regardless of their emotions. They can do it because they can disconnect from the everyday problems, hardships and other things which are not connected to the actual sense. The individual's conclusions are therefore not influenced by anything that can distort them. As a result, these people are often considered to be insensitive. As we are talking about a highly Life-supportive level, such people use their free energy and the fruit of their conclusions to make their environment better. Einstein was infinitely an idealist and humanist. He invested incredible efforts to set Humankind in a good direction. I read his writings that were not in connection with the science of physics when I was younger. Idealism and the desire radiated from them to lead Humankind's developmental tendencies in a better direction. If Humankind had listened to him better, the world would not have been in that situation, but instead at that time world was covered with wars and overheated nationalism. Einstein had no chance against so much negative energy. These negative energies have now been reduced, and the global communication systems can help good thoughts to spread. Namely, there are bigger chances to make the world better than there were in Einstein's time. It would be great to have Einstein among us!

Understanding is the main feeling at this level, as at this level we are able to disregard our own problems and mindset, and we are able to fully understand others. As a result, people living at this level can reveal the world's processes that are incomprehensible to others. It is not by accident that most of the Nobel laureate researchers, inventors, famous thinkers and philosophers live at this level. Their mindset is about significance. They think they received their abilities and skills to be able to show the right way to society with their significant results and by showing an example. If they believe in God, they imagine God as an infinitely wise entity who manages world fate with correct and objective decisions.

I had the opinion for a long time that there is no perfect objectivity since everybody values the world through their own distorted filter. Now I know there is perfect objectivity, but it is given to only a few people. People at this level have the luck to experience it every day but only in those fields they have enough experience in.

Love

I sometimes spend short periods of time at this level, so I already have experiences but I am really far from it on a daily basis, even while I am writing these lines. However, I am really striving with several methods to develop towards this wonderful world. People at this level understand that selfless love breaks through everything and it is the main power of the Universe. This understanding happens in the individual in a way that they fully experience love in everything and everywhere. The individual at this level does not need a partner or family to experience love since they feel this emotion towards every living creature. Those who reach this level are able to feel the all the love of the Universe. They do not need to get love feedback from other individuals to feel they can be loved. Mother Theresa lived at this level. Such People radiate so much love that it sticks to their environment and wonderful feelings develop in us if we are close to them. The person is in such harmony themselves that you feel it instinctively if you are close to them. A sense of unity with all living creatures appears in them quite frequently. This is why these people cannot hurt even a fly; they respect and love every living creature infinitely.

It is really the level of high-level spiritual leaders. Its energetic value is 500. Only a few people can reach it in their life. I know a lady whom I am pretty sure lives at this level. She adopts from hospitals children who were born with serious disabilities and abandoned by their parents. The children are dying, the pain they feel because they are abandoned is enough for that in itself, not to mention their disabilities which the doctors label as incurable. She brings up self-dependent happy adults from these little babies close to death with the power of selfless love. She cures them with the power of love while according to medical science it doesn't exist. This lady is infinitely peaceful, harmonious, patient and selfless. I didn't understand before how she can be like that or where she was getting the enormous amount of spiritual energy from. She brings up 10 to 12 children alone, only from donations, at the age of 65. Now I can understand the emotions of her existence and I also understand where she gets all the energy from. Even though I am not at her level, I can feel and understand how her spirit works. I know that this level exists and it is not fake, but a true and wonderful reality.

However, the average person is really far from it but in the distant future, when the average person will be at this level, Humankind will be living in world peace and perfect harmony with Nature. There will be no selfishness, aggression, or environment destruction.

The main emotion at this level is devotion, which the individual feels towards the wonder of Life (according to your religion it can be replaced by God, Buddha, the Almighty, etc.). The individual is infinitely benevolent and they try to declare it everywhere they find someone who is interested. Jesus' words make sense on this level: 'return good for evil'. People can see it clearly on this level that if someone hurts others, the biggest problem they have is with themselves. Instead of opposition, love is the only solution with which the suffering person can be set in the direction of healing.

If the people living at this level are believers, they experience God as an infinitely merciful entity. The positive level of consciousness of these people is so high that they can balance alone the negative level of hundreds of thousands of people. This is why we like to be close to these people since they make us more elevated and harmonious. This is also the reason why followers quickly gather around these people and spread their name if people are open to it. As humbleness is basic on this level, I am convinced there are many people in society living at this level, withdrawn from the world, without being aware of what power they radiate and with which they slow down the destruction of Earth dramatically. These people do it of course selflessly and unconsciously since they don't know about the scientific discoveries of kinesiology described here.

Joy and Peace

I am writing about Joy (energetic value 540) and Peace (energetic value 600) in one subchapter. Obviously, all of us have already felt these feelings, and every person with realistic values likes to be in that spiritual state. However, peace and joy also have levels. It can often happen that we feel ourselves peaceful but our leg is shaking or it turns out from a survey on the internet that we are actually a little bit stressed. This is because our brain compares us to ourselves. When we feel ourselves peaceful or happy, we feel it compared to our general irritation level. But most of us don't know what full inner peace is like, as we have never been there. You should imagine it as it was described at the level of Love in the previous chapter. Most people have already felt love with all their heart obviously, but hardly any of us feels perfectly selfless love, felt by every living creature, especially continuously. Those who feel it live at the level of Love.

Even fewer people live at the level of Joy and Peace than at the level of Love. The higher the level we discuss, the fewer people we're talking about. Those who live at these two levels mean only one person out of tens of millions. I regularly listen to Gunagriha's lectures on the internet. I am convinced he lives at least on this level. People living at the level of Joy experience the sense of their own wholeness. It means that they have accepted themselves so perfectly that they find nothing in themselves that cannot be loved. This must not be confused with conceit. At the same time, the inner spiritual attitude can be felt. Such people live the wholeness of Life.

The level of Peace is even higher, since the individual experiences the wonder of Life's perfection inside and outside. They reach this level regarding both self-acceptance and the acceptance of the world. People living at this level can neutralize the negative energy of hundreds of thousands of people living on Life-destructive levels. If such a person lives in a city, people are more peaceful there without knowing it's because this person is nearby. In that city there are fewer accidents and criminal activities.

People living at the level of Joy are steeped with the feeling of serenity. Such spiritual leaders always smile, and it's not a learned armor like with a lot of people, but the projection of the constant joy and serenity that live within them. If such a person looks into your eyes, you feel a high level of peace and harmony, your backbone starts to tingle and you will be impressed by the moment for hours.

People living at the level of Peace live in the feeling of being blessed. They experience life and themselves as blessed, and this huge peace which they experience in their spiritual safety is the strong basis of their spirit. This is the reason why there is no nervousness and stress, there is no fear or other forms of spiritual pain. Their mindset is about infinite goodwill and experiencing the wholeness of their life. If these people believe in God, those living at the level of Joy experience their unity with God, while those living on the level of Peace see God existing in everything. The individual will melt into the divine will and they cannot really be separated. These people are beyond the common practice of religion; they already live in the spiritual world which is the base and root of every religion. They do not learn the religion any more, but modify and shape it. They usually live withdrawn from the world since God-consciousness does not enable the traditional, ordinary way of life.

With all of their deeds they serve Life, so they do a lot to save Humankind and terrestrial Life. However, they are extremely shy so most of the time we don't know about them. A lot of people living at the level of Peace were canonized in the past or they founded new religious trends.

Perception at this level is like watching the world's happenings in slow motion in a highly transparent context where every happening makes sense. The consciousness does not try to create a concept to everything. The mind is always silent here; the perceiver becomes one with the perceived. 'Everything is connected with everything' makes sense on this level and becomes part of everyday perception. The individual lives in the perfect presence continuously which other people search for during meditations but can rarely find it for more than hours or minutes.

These levels also mean the world of wonders. People living at this level are able to do things which an average person can't even comprehend rationally. While some people perceive these things as wonders, skeptical people search for a rational explanation to prove that the wonder was a cheat or quackery. However, from among the spiritual leaders the excellent ones live on this level and the real spiritual healers live here who can heal by laying on of their hand. They can activate high-level spiritual energies which are beyond the current physical laws known today.

It frequently happens that people after experiencing near-death experiences get to this level. After they return to the living, they skip to this high, enlightened level. Most people are skeptical with them since **one of the cornerstones of narrow-mindedness is that we don't believe what we haven't experienced, as we know it doesn't exist.** This is why we feel an instinctive opposition to these people's revelations. I have already experienced so many things in my life that I had thought to be impossible that I learnt this is one of the biggest pieces of nonsense. In most cases it is unconscious and instinctive in people, since they hold on to their own experiences to be safe. They feel that if they accept it, their uncertainty would increase. But the reality is just the opposite.

Enlightenment

We have reached the highest level of existence, namely the level of Enlightenment. Its energetic value is between 700 and 1000.

At this level everything becomes perfect. The consciousness sees Life and the world as a shining wonder. The spirit is totally clear on this level; the consciousness does not link any thoughts, emotions or attributes to anything. People whose spirit is at this level live in the feeling of pure existence. You just are, as a part of a perfect system that we call Life, and in which you experience unity with everything else. It is such an uplifting and perfect feeling that it is hard to name it. The harmony building up from the unity of peace, joy and selfless love is about to overflow.

It is extremely rare: only one in hundreds of millions or in a billion is at this level permanently. At this level, existence itself is the meaning of Life. Existence is the aim. Realization itself is perfect as it is. There are no concrete aims or wishes; the sense of being steeps every moment. Here you simply do not need anything that common people need since existence itself is perfect as it is. The sense of time is lost but time makes no sense.

The people who live at the level of Enlightenment live in these feelings continuously. Nothing can make them leave this condition. The self is already part of the big unity. It is not by accident that Jesus did nothing to avoid crucifixion. Here is no good or bad; here is just being. People living at lower levels can reach the level of Enlightenment only for minutes, in general when everything is perfect around them and they can get into a properly deep meditative state. Meditators can search for this state throughout their lives without any results. By contrast, a person living at this level cannot be distracted from this state, even by being crucified. This is the huge difference between enlightenment and experiencing it only for minutes. There is such a huge distance between a common and an enlightened person that the average mindset cannot understand it.

Those people who lived at this level had an incredible impact on the world. The level of consciousness of such a person is able to compensate or neutralize the Life-destructive levels of several millions of people. Wonders become common at this level as the spirit is on an energy level that makes the individual able to do things which seem impossible by logical thinking. The founders of world religions and world-changing oracles lived at this level in the past, like Jesus, Buddha, Krishna, and Lao Ce. They are whose sentences will survive until the end of the world and will have an impact on Humankind. They are the people who are said to do wonders. Huge world religions were built on their sentences, however, people living at lower levels ruined and distorted a lot of these revelations by their comprehending and comments. This is why the church can never be as pure as the root from which it developed. Churches could maintain their pure characters only if there was always an enlightened spiritual leader heading it. Unfortunately, this has not been given to Humankind yet. However, it is also true that enlightened people only exist and they do not promote themselves. They do not want to be leaders of organizations. People just start to follow them at best. If we see this issue from this perspective, we will not need any churches in the far future. Church can be the crutch of the temporary period in case it can keep or regain its purity.

If you want to know more about this topic, I highly recommend David R. Hawkins *Power vs Force*, which was the basis of the previous chapters.